A JESUIT'S GUIDE TO THE
STARS

A JESUIT'S GUIDE TO THE

STARS

Exploring Wonder, Beauty, and Science

GUY CONSOLMAGNO, SJ

LOYOLAPRESS.
A JESUIT MINISTRY

LOYOLA PRESS.
A JESUIT MINISTRY

www.loyolapress.com

Copyright © 2025 Guy Consolmagno, SJ
All rights reserved.

For a comprehensive listing of editorial permissions, please refer to page 213.
The acknowledgments for all artwork can be found on pages 215–216.

To purchase copies of this book in bulk, please contact our Customer Service
team at (800) 621-1008.

ISBN: 978-0-8294-5573-1
Library of Congress Control Number: 2024939106

Published in Chicago, IL
Printed in China
24 25 26 27 28 29 30 31 32 33 Dream Colour 10 9 8 7 6 5 4 3 2 1

To the memory of Fr. Martin McCarthy, SJ (1923–2010),
who showed me the joy of being a Jesuit astronomer.

CONTENTS

Galaxy NGC 3628, imaged at the
Vatican Advanced Technology Telescope (VATT).

A NOTE TO MY READERS

THIS BOOK IS ABOUT JESUITS AND STARS. Lots of Jesuits—from historical figures to me, the author. Lots of stars—clusters and galaxies of stars. And around them, planets and asteroids and meteorites as well.

Too often we fail to realize our remarkable place in the universe. We get bogged down by personal crises and the problems of the world. Sometimes we fear that we must abandon either our religion or our science as we try to find our way out of despair. We let our curiosity about the universe go stale. Those of us with a scientific bent, like doctors and engineers, can lose sight of how our science is rooted in religion and its search for truth and beauty. Some people think that science and religion are at war.

This book shows how each illuminates the other.

All my life I have loved to look at the stars. But even more, I love how the stars themselves remind me of a universe so much larger than my own mundane concerns. I love how my knowledge of those stars reminds me of my own personal history, as well as my culture's history and my religion's history of pondering those stars. And I love how the beauty in the sky is but an invitation to the surprising beauty found in astronomy itself, in science, in how we come to understand what we see.

The stars are beautiful. It is a joy to contemplate them. They can be a place to find God. Finding God in the stars is a quest dating back to the Psalms and Scripture, and it is something that anyone can do even today. But it just might be helpful to have a guide to show the way.

BROTHER GUY
Vatican Observatory

St. Ignatius and the Stars, Collège Saint-Michel, Fribourg, Switzerland

INTRODUCTION

IN THE *AUTOBIOGRAPHY OF ST. IGNATIUS,* which Ignatius dictated to a scribe who recorded it in the third person, we read, "The greatest consolation he received was to look at the sky and the stars, which he often did and for a long time . . ." St. Ignatius of Loyola (1491–1556) was the founder of the Society of Jesus, commonly known as the Jesuits. His first companions, the first Jesuits, were a group of men who met at the University of Paris in the mid-sixteenth century. They developed a unity of hearts and minds from being students together; scholarship was an essential part of what formed them as a Society. Their scholarship, from the beginning, was at the highest academic level, and it included expertise in all the subjects of the university—not only theology and philosophy but also the subjects of the medieval quadrivium: mathematics, music, geometry, and astronomy.

As its name indicates, the Society of Jesus is centered on the person of Jesus. To Christians, Jesus is God's incarnation within our physical universe, and so it follows that the spirituality of the Society of Jesus is *incarnational*. What does that mean? Consider the famous Jesuit mantra, "to find God in all things." Note the emphasis on *things*: the created world. In that phrase, being close to the created world is taken to be a way of being close to the Creator. This incarnational spirituality is rooted in what we experience when we immerse our minds and our hearts into the study of the universe, when, like St. Ignatius, we spend time looking at the stars.

Throughout their history, Jesuits have had a special connection both to the beauty of the stars and to the study of astronomy. From the first modern map of the Moon (where two dozen craters were named for Jesuits—by the Jesuit who drew the map!)

to the latest understanding of the Big Bang, Jesuits have been found at the frontiers of astronomy. They have also been at the forefront of educating students and scientists about the wonder of the universe.

A Jesuit's Guide to the Stars is my way, as a Jesuit and an astronomer, of introducing a wider audience to the nighttime sky. It includes stories about the people who have opened up our understanding of the universe, connections between Ignatian spirituality and astronomy, and activities that can help make that universe feel as familiar to you as your own backyard—because it is your backyard!

Most people experience astronomy simply by looking at the stars, or at least by looking at images of things in space as our telescopes and our spacecraft see them. Thus, large parts of this book are photo essays, beginning with Scripture itself. In keeping with the book's Jesuit theme, virtually all the astronomical images were made by Jesuits or astronomers associated with the Jesuits at the Vatican Observatory, using our various telescopes. (The exceptions are those images that come from spacecraft; the Vatican doesn't have its own space program—yet!) A number of sky images (I took those) are included as well; after all, the sky itself was also an inspiration to St. Ignatius.

And this is also but one Jesuit's guide: mine. Woven through these photo essays is my personal story. A lot of it is my take on what the universe looks like to me . . . as an astronomer, as a Jesuit, and as a kid from Detroit who somehow wound up at one time or another on every continent on Earth, pursuing the knowledge of what's above Earth and how it affects us all down here.

My own formation as a Jesuit began with the Spiritual Exercises. My life as a Jesuit gave me the chance to pursue my passion both as a stargazer and as a meteorite scientist. My studies with the Jesuits introduced me to philosophy, history, and poetry, and they gave me the chance to live and work with astronomers who ponder the very

questions of who we are and where we come from. And my work at the Vatican has made me intimately familiar with the vision of that wonderful fellow Jesuit, Pope Francis, and how his encyclical *Laudato si´* defines our place in this universe.

The Moon belongs to everyone. We all live under the same sky. But encountering this universe with the mindset of a Jesuit means going beyond just looking up and thinking, *Oh, wow, look at the Moon.* It is experiencing everything in a unity of hearts and minds.

Engaging the universe with the heart means not only appreciating its beauty, but also recognizing the love that lies behind that beauty and feeling the joy that is the sure sign of the presence of God in his creation. Engaging the universe with the mind means not only puzzling out its scientific mysteries but also remembering the rich human history behind how we understand the sky . . . and our own personal histories, as well.

A Jesuit's eye on the sky means beholding it with both nostalgia and amazement, familiarity and mystery, awe and joy—in everything. That's why they call it the Universe. It is the "all things" where we find God.

View of Earth over Moon's horizon taken from Apollo 11 spacecraft.

The Andromeda Galaxy, imaged by Claudio Costa with the Vatican Observatory's wide-field camera.

CHAPTER ONE
STARS IN SCRIPTURE

PEOPLE OFTEN ASK ME IF THE STARS are found in the Bible. Of course they are! What is fascinating to me is *how* the Bible talks about stars, and how they are used as pointers to the One who made them. I don't pretend that what follows is a complete inventory of stars in Scripture, but it certainly is a representative sampling.

One way to look at the bright things in the sky is to see them the way that all ancient peoples used the stars, as timekeepers. Following the stars' rhythms is useful if you want to know when to plant or reap, when to expect high tides or bright moonlit nights, how to keep track of the year and the years:

> The sun, when it appears, proclaims as it rises
> what a marvelous instrument it is, the work of the Most High. . . .
> Great is the Lord who made it;
> at his orders it hurries on its course.
>
> It is the moon that marks the changing seasons,
> governing the times, an everlasting sign.
> From the moon comes the sign for festal days,
> a light that wanes when it completes its course.
> The new moon, as its name suggests, renews itself;
> how marvelous it is in this change,
> a beacon to the hosts on high,
> shining in the vault of the heavens!
>
> The glory of the stars is the beauty of heaven,
> a glittering array in the heights of the Lord.
> On the orders of the Holy One they stand in their appointed places;
> they never relax in their watches.
>
> SIRACH 43:2, 5-10

The waning Moon setting over the desert at the Redemptorist Renewal Center north of Tucson, Arizona.

But the stars are more than just utilitarian timekeepers. Stars are also signs of God's omnipotence, infinity, and everlasting love for his people:

> O give thanks to the LORD, for he is good,
> for his steadfast love endures forever; . . .
>
> who alone does great wonders,
> for his steadfast love endures forever;
> who by understanding made the heavens,
> for his steadfast love endures forever;
> who spread out the earth on the waters,
> for his steadfast love endures forever;
> who made the great lights,
> for his steadfast love endures forever;
> the sun to rule over the day,
> for his steadfast love endures forever;
> the moon and stars to rule over the night,
> for his steadfast love endures forever.
>
> PSALM 136:2, 4–9

Indeed, in Scripture we hear about the stars themselves giving praise and adoration!

> Then the LORD answered Job out of the whirlwind: . . .
> "Where were you when I laid the foundation of the earth?
>   Tell me, if you have understanding.
> Who determined its measurements—surely you know!
>   Or who stretched the line upon it?
> On what were its bases sunk,
> or who laid its cornerstone
> when the morning stars sang together
>   and all the heavenly beings shouted for joy?"
>
> JOB 38:1, 4-7

And again:

> O Israel, how great is the house of God,
>   how vast the territory that he possesses!
> It is great and has no bounds;
>   it is high and immeasurable . . .
> the one who sends forth the light, and it goes;
>   he called it, and it obeyed him, trembling;
> the stars shone in their watches, and were glad;
>   he called them, and they said, "Here we are!"
>   They shone with gladness for him who made them.
> This is our God;
>   no other can be compared to him.
>
> BARUCH 3:24-25, 33-35

Horsehead Nebula, imaged at the Vatican Advanced Technology Telescope on Mt. Graham, Arizona.

The Bubble Nebula, NGC 7635, imaged at the Vatican Advanced Technology Telescope. The "bubble" is formed by the hot wind emitted from a star inside a cloud of gas and dust located about 10,000 light years away.

Stars tell us about the nature of God, who created them. He loves beauty and elegance. He cares for them, naming them and calling them by name:

> He determines the number of the stars;
> he gives to all of them their names.
> Great is our Lord, and abundant in power;
> his understanding is beyond measure.
>
> PSALM 147:4-5

And that is because God is bigger than any of them and all of them. Stars are God's creation, designed for delight and beauty, not objects to be worshiped, as the neighbors of Israel were tempted to do:

> God made the two great lights—the greater light to rule the day and the lesser light to rule the night—and the stars. God set them in the dome of the sky to give light upon the earth, to rule over the day and over the night, and to separate the light from the darkness. And God saw that it was good. And there was evening and there was morning, the fourth day.
>
> GENESIS 1:16-19

Total solar eclipse, 2017; image by Claudio Costa, Vatican Observatory.

In the Genesis teaching, God begins by creating light, so that nothing he does is hidden, nothing is in darkness, nothing is removed from our sight. To quote a bad sci-fi movie trope, there are not "things man was not meant to know!" But as Pope Benedict pointed out in his homily on Holy Saturday, April 7, 2012: "The sun and the moon are created only on the fourth day. The creation account calls them lights, set by God in the firmament of heaven. In this way [the author of Genesis] deliberately takes away the divine character that the great [pagan] religions had assigned to them. No, they are not gods. They are shining bodies created by the one God. But they are preceded by the light through which God's glory is reflected in the essence of the created being."

> And when you look up to the heavens and see the sun, the moon, and the stars,
> all the host of heaven, do not be led astray and bow down to them and serve
> them, things that the LORD your God has allotted to all the peoples everywhere
> under heaven.
>
> DEUTERONOMY 4:19

Helix Nebula, a planetary nebula, imaged at the Vatican Advanced Technology Telescope.

It is a sign of wisdom to recognize that the glorious things in the sky are not themselves gods but rather things made to direct us to the one God, their Creator.

> For all people who were ignorant of God were foolish by nature;
> and they were unable from the good things that are seen to know the one who exists, . . .
> but they supposed that . . . the circle of the stars . . .
> or the luminaries of heaven were the gods that rule the world.
> If through delight in the beauty of these things people assumed them to be gods,
> let them know how much better than these is their Lord,
> for the author of beauty created them.
> And if people were amazed at their power and working,
> let them perceive from them
> how much more powerful is the one who formed them.
> For from the greatness and beauty of created things
> comes a corresponding perception of their Creator.
>
> WISDOM 13:1-5

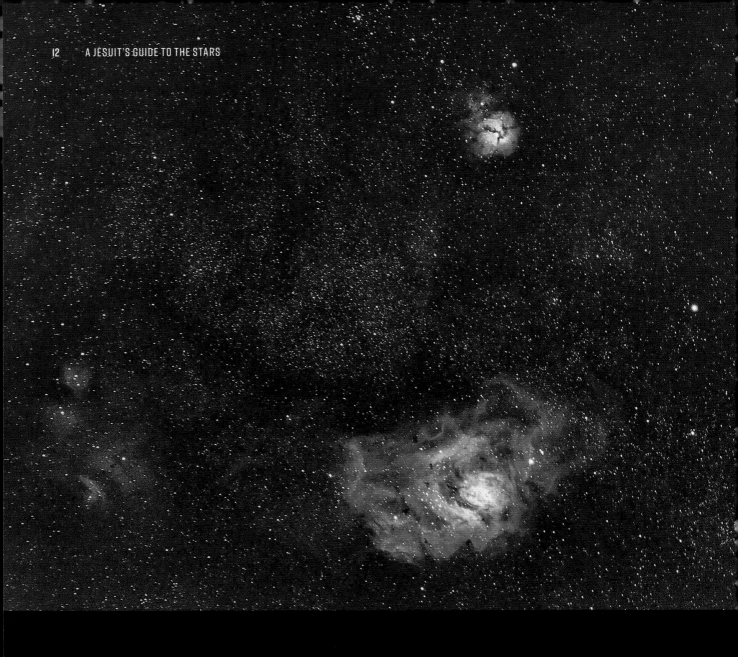

Star-formation regions in the Milky Way: the Trifid Nebula (top center), the Lagoon Nebula (bottom center), and Sharpless 221 (SH2-221) (bottom left); imaged with the Vatican wide-field camera by Claudio Costa.

Even if stars are not to be worshiped in themselves, Scripture tells us that they are an example of beauty exceeded only by the wisdom of God. And even more than that, the wisdom that gives us the possibility of understanding the stars is itself an incredible, beautiful gift of God:

> May God grant me to speak with judgment,
> and to have thoughts worthy of what I have received;
> for he is the guide even of wisdom
> and the corrector of the wise. . . .
> For it is he who gave me unerring knowledge of what exists,
> to know the structure of the world and the activity of the elements;
> the beginning and end and middle of times,
> the alternations of the solstices and the changes of the seasons,
> the cycles of the year and the constellations of the stars,
> the natures of animals and the tempers of wild animals,
> the powers of spirits and the thoughts of human beings,
> the varieties of plants and the virtues of roots;
> I learned both what is secret and what is manifest,
> for wisdom, the fashioner of all things, taught me. . . .
>
> She is more beautiful than the sun,
> and excels every constellation of the stars.
> Compared with the light she is found to be superior,
> for it is succeeded by the night,
> but against wisdom evil does not prevail.

WISDOM 7:15, 17–22, 29–30

Most surprisingly, in this fabulous creation of stars and sky, we humans have our place:

> O LORD, our Sovereign,
>   how majestic is your name in all the earth!
>
> You have set your glory above the heavens. . . .
> When I look at your heavens, the work of your fingers,
>   the moon and the stars that you have established;
> what are human beings that you are mindful of them,
>   mortals that you care for them?
>
> Yet you have made them a little lower than God,
>   and crowned them with glory and honor.
> You have given them dominion over the works of your hands;
>   you have put all things under their feet,
> all sheep and oxen,
>   and also the beasts of the field,
> the birds of the air, and the fish of the sea,
>   whatever passes along the paths of the seas.
>
> O LORD, our Sovereign,
>   how majestic is your name in all the earth!
>
> PSALM 8:1-9

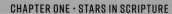

Sunset, the Pontifical Gardens of Castel Gandolfo, from the roof of the Vatican Observatory headquarters.

Some people are daunted by the sheer scale of the universe; they think that modern astronomy has reduced the importance of human beings. But 2,700 years ago, the psalmist understood that we can seem to be nothing in the face of all those stars, and the psalmist knew the answer to that quandary. It is not our stature, but our Creator, who has made us important.

There are two ways to look at this immensity. We can ask, *If the universe is so big and I am so small, how can God notice me?* Or we can say, *If the universe is so big and I am so small, how immense God must be not only to have been responsible for that universe but also to notice me, lost in the midst of it all.*

Three of the five galaxies known as Stephan's Quintet, as imaged at the Vatican Advanced Technology Telescope. This grouping of galaxies is found in the constellation Pegasus.

Just as stars are grouped together into galaxies, so galaxies themselves form clusters of galaxies. This cluster, called Abell 397, has at least thirty-five member galaxies; it is located in the direction of the constellation Aries. Vatican Advanced Technology Telescope image by Matt Nelson.

Jupiter and its four brightest moons; Vatican Advanced Technology Telescope image by Br. Jonathan Stott, SJ.

CHAPTER TWO
AN ASTRONOMER'S ROAD TO THE JESUITS

IT WAS A FINE SUMMER'S EVENING. After spending all day in a lecture hall listening to a parade of scientists giving reports of their research, it felt wonderful to step outdoors.

I was just twenty-one years old, attending my very first scientific meeting. The topic was planetary satellites. Several dozen scientists from around the world had gathered at Cornell University in upstate New York to describe what we knew about the moons that orbit the planets of our solar system.

This was back in 1974. Planetary science was still a very young field, so young, in fact, that I suspect that this was the very first international meeting that dealt solely with planetary satellites. Certainly, it was the first such meeting since the dawning of the space age, which was itself not even twenty years old at that time. We had no good spacecraft images yet of any moon except our own Moon, but we knew that the Voyager spacecraft, just launched, would soon be sending us actual data about the bodies that were in orbit around the outer planets. We wanted to be ready to know which questions to ask, what to expect, what to be surprised by.

In 1609, Galileo first spotted little dots of light moving back and forth near Jupiter. He recognized them as bodies moving about Jupiter. His realization, in fact, was more important than just seeing the dots of light. Sooner or later other people would have seen those dots, but it took Galileo's genius to understand that he was seeing *moons* orbiting Jupiter, just like our Moon orbits Earth.

Even some 350 years later, we still didn't know much more about those moons than what Galileo had deduced. The only additions to our knowledge came in the nineteenth century, when astronomers measured the moons' size and intrinsic brightness (much shinier than rock) and density (much less than rock). By the 1920s it dawned on a few

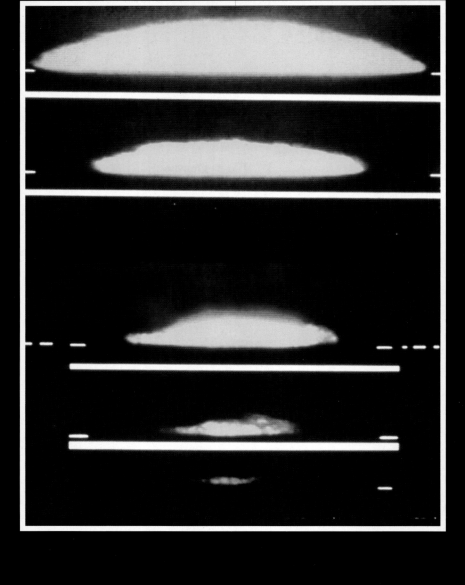

First published photograph (1958) of the "green flash," a rare atmospheric phenomenon in which the last sliver of the setting Sun turns green. Photos by Karl Treusch, SJ, Vatican Observatory.

people that this meant they were probably rich in water, in the form of ice. Only three years before the 1974 meeting I was attending, the characteristic infrared spectrum of ice was detected in the light reflected from the moons' surfaces.

My undergraduate research at the Massachusetts Institute of Technology (MIT) was to write computer models to predict how icy moons like these might warm up and perhaps even melt the ice inside of them. I spent more than a year struggling to write code that would run on what were, at that time, some of the fastest computers in the world. Then I had to translate the numbers that came out of the computer into an understanding of just what might be happening, geologically, inside these distant icy worlds.

> As a kid I had looked at it in my own little telescope, and I had seen the same little specks of light beside it that Galileo had seen.

My advisor was impressed with what I had accomplished; my models *almost* worked. He arranged for me to complete my work as a master's thesis, taking one more year before heading off to do a doctoral degree. This meeting at Cornell was my first introduction to the wider world of planetary science, a place where I could compare what I was doing with what else was happening in the field.

And so I stepped outside that evening, my mind reeling with thoughts of orbits and impacts, exotic ices and primitive minerals. And then I looked up.

In the sky overhead was a bright dot of light.

Now, when I was growing up in Michigan, my father would point out the stars and planets to me. So I knew that dot. That dot was Jupiter. As a kid I had looked at it in my own little telescope, and I had seen the same little specks of light beside it that Galileo had seen. Those pindots were the very moons that we scientists had been talking about all day.

I realized at that moment when I stepped outside how remarkable it is to be a planetary scientist. I can go outside on any clear night and look up to see the very thing that I am studying. But I also experienced at that instant a wonderful, if not jarring, disconnect as I tried to relate the figures and computer codes spinning through my brain with the tiny bit of light hitting my eye.

I imagine that's how theologians must feel when, after a day of erudite study of the nature of God, they retreat to a chapel for a moment of prayer.

The reality of the stars and planets overhead is essential to my science, just as the reality of the God we encounter in prayer is essential to one who participates in the oldest science, theology. But it takes an enormous stretch of the human brain to connect the immediate experience of stars, or God, with the not-quite-certain aggregation of speculation that the human race has learned to attach to that experience.

Both scientists and theologians must remove distractions—random city lights that pollute the night sky, random concerns about lunch and bills and assignments to grade that fog our minds—to see the subtle, faint glow of a distant moon. To hear the still, soft whisper of God.

How do we do it?

We do it by working within a community of scholars, learning what those before us have discovered, hearing about what others may have experienced in some fleeting moment of stillness or clarity, testing our intuitions against those of friends and fellow seekers. This is what lets me appreciate the significance of my experiences at the telescope—or in the chapel. Only then can the sheer reality of that light hitting my eye, hitting my soul, confirm everything else that I do.

No wonder Saint Francis saw in the Sun and the Moon, the stars and the planets, places that give praise to God just by existing. They direct us to the Source of their existence.

The way I got into the science of studying planets was not quite so exalted. When I was young, the Moon and stars and planets were not things that would lead me to God. They were places where people could have adventures.

I had been a fan of science fiction books ever since I learned to read. In the adventures of

The Triangulum Galaxy; Vatican Advanced Technology Telescope image.

heroes jetting around the solar system, I experienced a sense of wonder, letting my imagination be led to places no one had ever seen. These were not "worlds unimaginable." They were worlds very much imaginable—worlds that in fact had been imagined, worlds that could be visited only with the imagination.

When I first left my home in Michigan for university studies in Boston, I had too many ideas about which field I wanted to pursue. What did I want to be when I grew up? A journalist like my father? A lawyer like my grandfather? A priest like the Jesuits who had taught me in high school? I enrolled in the Jesuit university in Boston, Boston College, because it offered good programs in all those topics and because I would remain close to the Jesuits. I hoped that once I was there, my future would come into focus.

Back then, one of my dreams was to write stories like those I had read as a child, and so I took a course on creative writing. The young Jesuit scholastic who taught me said that to learn to write good fiction, I should read good fiction: stories that were well written. Since I was fascinated by stories from my childhood, he suggested *The Chronicles of Narnia* by C. S. Lewis.

Oddly, although they were published during my childhood, I had never come across them when I was growing up. Where would I find them now, as a college student? I mentioned my quest to my high school friend Mike Timmreck, who was also studying in Boston, at MIT, which was a fifty-minute trolley ride from Boston College. He guided me to the science fiction library maintained on campus by the MIT Science Fiction Society. It was, in fact, the largest open collection of science fiction in the world.

In that library, all the joys of my youthful infatuation with science fiction came rushing back . . . so strongly that I abandoned thoughts of law or journalism or priesthood. *How could I get closer to living with these adventures?* I asked myself.

Reading the greats in a private retreat space. Lapeer, Michigan, 1971.

I plotted my transfer to MIT. I filled out the forms, went for the interviews. Predictably, MIT wanted to know what I would be studying if I came there. I looked over the choice of concentrations and found one called Earth and Planetary Sciences. Planets! That's where people had adventures! To realize that there were actual worlds out there, someplace in space, where someone (if not me, then some future person like me) might actually have adventures—that captured my imagination.

Of course, I didn't say that I wanted to come there to dream up science fiction stories. I said that I wanted to be a science journalist. On paper it must have sounded reasonable; back in Michigan I had been editor of my high school paper, and I was working as a summer intern at the *County Press,* in Lapeer, Michigan. To be honest, though, I didn't think MIT would accept me. MIT is one of the world's premier science and technology universities, and it turns down nine out of ten students who apply. My studies up to that time had been in classics and literature, not in science. But I guess they figured the world needed more science journalists. For whatever reason, they let me in.

I will never forget my shock, opening up the acceptance letter and reading it in disbelief. In fact, I still have that very letter in a frame on my wall as a reminder that miracles do occur. But my joy was soon followed by dismay when I discovered that the department that had accepted me, Earth and Planetary Sciences, was actually not astronomy but . . . *geology.* Rocks? I wanted to have adventures! What could be more boring than studying a bunch of rocks?

Still, if geology was my ticket to going to school near that science fiction library, I guessed that I could pretend to be interested in rocks, for a couple of years anyway.

The first semester at MIT taught me many surprising things. I learned that if you skipped out on lectures, nobody noticed—until you failed the exams. I learned that the basic physics course everyone was required to take was utterly baffling to me. But I also learned that rocks could actually be fascinating.

And I learned the exhilaration of just being at MIT, someplace special and wonderful, where *real things* were happening all around me. The sheer joy of walking the halls of The Institute (and its basements, which were particularly fascinating late at night)—it felt like I was living in a science fiction story.

A major crisis came in January 1972. At MIT, January is set aside as a time to explore topics that might be outside the normal syllabus. It is also a time when you can catch up on prerequisite coursework that you might have missed. Since I had arrived at MIT a year later than the rest of my graduating class, to stay on track with them, I needed to fit the entire second semester of introductory physics into one month of intensive study.

> I learned that if you skipped out on lectures, nobody noticed—until you failed the exams.

How could I be expected to do that? I'd barely passed the final exam of first semester mechanics (I had to take it twice), and I understood none of it. The January session on electricity and magnetism was worse. The course used a textbook that must have been written by a sadist, someone who hated students, someone who wanted to flaunt his knowledge and make me feel stupid. It was infuriating. I couldn't possibly solve the problems at the end of each chapter. I didn't even know where to begin.

Finally, in desperation, I took one of the assigned problems and just arranged the ingredients so that all the different units (meters here, grams there, seconds over there) matched what I knew the answer was supposed to look like. I took my result to the class tutor. He nodded and said, "Yes, that's how you do that."

What? I thought I was just faking it.

"It's called dimensional analysis. It is how all physicists approach a new problem. It's how we see the way the different elements of a problem fit together."

My mind was spinning. Could it be that easy? I went back to the problems from the earlier chapters. They all made sense now. In fact, they seemed ridiculously easy.

I revisited the problems from the mechanics course, the one I had nearly failed. The answers had become obvious too. In fact, they seemed so simple that I didn't feel any sort of accomplishment in getting them right. The answers were so plain that anyone should have been able to see them.

(Six years later, and with a doctorate in hand, I began postdoctoral research at Harvard University. One of the professors I worked with, Ed Purcell, was a Nobel laureate in physics. He had discovered the physical principle behind the MRI scanners you find in hospitals today. Dr. Purcell was friendly, funny, charming, and a delight to work with. And he was the guy who had written that textbook I had cursed so strongly!)

My personal physics breakthrough echoed something that had happened just a year earlier. While still at Boston College, I had gotten word that my mother was diagnosed with breast cancer. It had scared me then like nothing I had ever experienced before. She would be going in for an operation. All I could do was pray.

And in my prayer two remarkable things happened. First, I felt an enormous and unexpected sense of peace. Second, I felt a connection to all the other people in the world facing crises and pain and fear, in what suddenly I recognized as the "Mystical Body of Christ." *Oh!* I thought to myself. *The Mystical Body of Christ!* I learned that phrase when I was younger. In that moment I knew what it meant.

My experience of God then was my experience of physics a year later. What were once empty words had suddenly become a reality. And it was a reality that I could express only by using the same words that had meant nothing to me until I experienced their meaning. To learn, I had to have the experience.

Incidentally, my mother lived to be a rousing ninety-seven years old.

In my second year at MIT, I was finally able to take a basic course in planetary physics and chemistry. The course was team taught by two professors who would become longtime mentors and collaborators, John Lewis and Irwin Shapiro.

I remember, even more than the science they taught, the enthusiasm with which they taught it. To them, science itself was a grand adventure, more thrilling than any adventure story that I could have found in a science fiction library.

When I was a kid, science fiction stories had told me that the planet Mercury kept one face to the Sun, just like Earth's Moon always faces Earth. And since Mercury is so close to the Sun, the sunlit side was naturally baking hot while the far side of Mercury never saw sunlight and so was freezing cold. Except, it turns out, that's not true. Dr. Shapiro was a member of the team that had bounced radar waves off the surface of Mercury and found that while it took eighty-eight Earth days to go around the Sun, Mercury spun on its own axis once every fifty-nine days. Dr. Shapiro also worked out the physics and mathematics of why it was inevitable that Mercury would move and spin at just that rate.

When I was a kid, science fiction stories had told me that Venus was covered with thick clouds covering up deep jungles. Except, it turns out, that's not true. Dr. Lewis took the observations from the first spacecraft to visit Venus and worked out the chemistry of a planet whose atmosphere was rich in carbon dioxide, the greenhouse gas with which we are all too familiar here on Earth. As a result, its atmosphere is hundreds of degrees hotter, and a hundred times denser, than Earth's. Chemical reactions between the carbon dioxide in that atmosphere and the rocks on the surface produce the clouds we observe on Venus. These were clouds not of water, but of sulfuric acid.

My science fiction stories had described planets that were just extreme versions of what we knew on Earth. Instead, reality was far beyond anything we could extrapolate from our own experience. That made me question just how well I could possibly imagine what it would really be like to experience eternal life or heaven or standing in the presence of God.

It is the real Sun and Moon and stars, not my pale imagination of them, that give praise to their Creator.

By 1978, four years after that meeting at Cornell, I had earned my doctoral degree in planetary sciences from the University of Arizona. Returning to Boston for postdoctoral studies, I spent two years at the Harvard College Observatory, then shifted back to MIT. By the time I was thirty years old, I was completing the final year of a three-year appointment as a researcher and lecturer at MIT.

I was in an odd place in my career. I was a researcher at two of the top universities in the world, but I was at the absolute bottom rung

The author with his parents, Patricia and Joseph Consolmagno, at MIT's graduation ceremony, 1974.

of the academic ladder, with no likelihood of climbing any higher. After five years of looking, I hadn't found any school eager to make me a professor. And I was thirty years old. *Old!*

My age weighed heavily on me. So did my conscience. I was a product of a Jesuit education. I was looking for "the *magis*"—I was wanting to do more with my life. I had even thought of being a priest. With the weight of thirty years of Catholic guilt on my chest, I would lie awake in bed at three in the morning wondering, *Why am I worrying about the moons of Jupiter when there are people starving in the world?*

The more I asked that question, the deeper the silence I heard in reply. Why didn't God answer me? Acting as much out of frustration as with reason, I threw this in God's face: "You duped me into going to MIT, and I let myself be duped. So I became a scientist, and now I have a meaningless life and no job prospects." *I'll quit science,* I decided. *I'll turn my back on my pointless life and my pointless career, and I'll join the Peace Corps. I'll go to Africa. That'll show God!*

The author at the equator on the road to Eldoret, Kenya, 1984.

I asked my sister about the Peace Corps; many years earlier, she had served with them in Africa. She warned me: "You know, with your degree, you'll probably be assigned to the university with a nice apartment in the capital, and all the other volunteers will stay with you when they come to town." That was not what I wanted. I wanted what I thought was the real thing, living in the remote countryside and helping people who were starving. But I was tired of trying to plan out my own career. I would go wherever they sent me.

The author with three students at the Starehe Boys' Centre and School, Nairobi, Kenya; 1984.

They sent me to Kenya. First, they trained me to teach in a small village school, like I had hoped; but when my assignment came, it was to teach physics at the best high school in the country, the Starehe Boys' Centre and School, in a lab equipped with computers and lasers. (This was 1983, when few schools even in America had that kind of equipment.) I spent one term at Starehe before I was reassigned—to teach at the University of Nairobi, and with a nice apartment in Nairobi. All my fellow volunteers could stay with me when they came to town.

My sister had warned me.

What's more, I realized that even when I had tried to quit science, God wouldn't let me.

Since my volunteer friends were visiting me in Nairobi, I returned the favor. I spent almost every weekend going up-country to their remote villages, seeing what "real" Africa looked like. I earned my keep by giving talks at their schools about the stars and the planets and what the space program was teaching us about them. I brought a little telescope with me, and after dark everyone in the village would line up to see for themselves the craters on the Moon and the rings of Saturn.

> What's more, I realized that even when I had tried to quit science, God wouldn't let me.

Not surprisingly, the reaction of the people in those remote villages was exactly the same as the reaction I would get back home in Michigan when I gave my talks and set out my telescope. I don't know anyone who has ever seen the rings of Saturn in a little telescope and not gasped with surprise and awe. I still do.

Meanwhile, back in Nairobi, I discovered that I loved teaching physics. I was as excited to lecture as I had been to learn. My undergraduates were enthusiastic about my introductory physics lectures (the very classes I had once almost failed), and the graduate students were eager to learn solid-state physics, electrodynamics, and astrophysics.

At one level, I understood why the students were so hardworking. Kenya had the highest birth rate in the world. Their schools were overflowing. The need for teachers was never ending. My undergraduates were guaranteed jobs as teachers, and my graduate students had jobs waiting for them at the Kenya Science Teachers College, where they would teach the teachers. In that sense, I suppose, by teaching the teachers who would teach the teachers, I was helping develop Kenya into a nation that could teach, and feed, itself.

Even so, this did not explain the joy that my students had in learning astrophysics. And it certainly didn't explain the folks in villages who lined up every weekend to look at the moons of Jupiter.

I had a very clever cat in those days that was much better at being a cat than I could ever be. But my cat never wanted to look through my telescope. Looking through a telescope and being curious about the universe is something that is peculiarly human. And indeed, if you don't give people the chance to engage their curiosity this way, then you are denying them their humanity. People were starving in Kenya, yes. Many went to sleep hungry. But a lot of them also lacked a chance to nourish their souls on the same intellectual food that we in the developed world are able to feast on. *We do not live by bread alone*—that echoed in my mind. There are so many kinds of hunger.

Why do we hunger to know more about the Sun and the Moon and the stars? Why do we look with amazement at the night sky? What is it we are hungry for?

The lights in the sky lead us to the Creator. In Baruch 3:34, the prophet Baruch describes a time when "the stars shone in their watches, and were glad; he called them and they said, 'Here we are!' They shone with gladness for him who made them."

Can inanimate objects give praise to God? I'm rather fond of inanimate objects, to be sure; I loved studying rocks. I have found that rocks can be easier to get along with than people can be. Certainly rocks are easier to understand; rocks don't change their minds or ask if we can be just friends. I suppose that is one reason I became a scientist instead of a priest.

In Africa, I learned why astronomy matters to everyone, even people in remote villages. In the process, I saw how balls of gas and ice can give praise. It happens when people are attached to those objects.

When I see the icy moons through my small telescope, I remember my father, who first pointed out to me that little dot of light in the sky. I remember Professor John Lewis, who directed my research into those moons. I remember my friend Dan Davis, who helped me buy a little telescope and taught me how to use it before I left for Africa.

When I look at the rings of Saturn, I can also see in my memory the countless students who have used my little telescope and whom I have watched as their eyes grew as big as those rings the first time they saw them for themselves.

Of course, I also remember the kids who put their thumbs on my telescope's lenses, leaving their indelible mark. I remember the political fights I had with other scientists about what the icy moons were really made of, and about who thought of that first, and who deserved the grant money to study the moons some more. The Jesuit poet Gerard Manley Hopkins described it well in his poem "God's Grandeur": "All is seared with trade; bleared, smeared with toil; And wears man's smudge and shares man's smell."

> But when we go to the stars, we will also bring with us joy and love.

When humankind goes to the stars, as my beloved science fiction books describe, we will also bring with us our cynicism and the smell of stale beer. This is inevitable. We are sinners; we are fallen creatures. It is important to remember that we lost paradise. Heaven is not to be found at the end of any spaceship journey.

But when we go to the stars, we will also bring with us joy and love.

Stars are glorious lights in the sky, a source of joy that the prophet Baruch saw as giving praise to God.

For millennia, people have described the arrangement of the stars we see in the sky by grouping them into pictures we call *constellations.* Of course, just because we see a group of stars together in our sky doesn't mean that they are necessarily next to each other in space. Some stars might be much closer to us, and others very far away. But these constellations do three wonderful things for us.

Open Cluster NGC 2158; Vatican Advanced Technology Telescope image.

First, constellations are a fun way to remember and keep track of the stars. When I go out at night I can immediately tell Rigel from Regulus by recognizing which constellation I am looking at.

Second, constellations connect me with my ancestors, from my father who taught me the stars back to the ancient Greeks and Babylonians who first wrote down the names and stories for those groupings. In fact, many of the constellations we use to organize the sky are both older than any written records and common to many diverse civilizations. For instance, the stars around the Big Dipper have been identified with a bear by Indo-European language speakers as well as by a number of Indigenous American cultures.

Third, the stars in their constellations can tell us where we are. My father knew the constellations because he was a navigator in B-17s during World War II. Using the stars, he led squadrons of airplanes from America across the Pacific Ocean to Hawaii for the Battle of Midway, and then across the Atlantic Ocean to England and over the battlefields of Europe.

My father used the stars to find targets during war. I used them to find myself during my service in the Peace Corps.

When I first arrived in Kenya, it soon struck me what a foolish thing I had done. I was not an adventurer; I was merely someone who liked to read about adventures. I didn't actually want to fly in space; I just wanted to daydream about it. I was a stay-at-home person, but I was far away from home. What was I doing there?

The stress of being so far from anyone I knew, so far from anything that I knew, grew on me day by day. The food was strange, the language was strange, and the very air was full of fragrances I did not recognize. And even just walking down the street, I realized that everyone else around me had healthy-looking black skin, while my white skin looked pale and sickly. I did not belong.

> My father used the stars to find targets during war. I used them to find myself during my service in the Peace Corps.

So finally, one evening, I made a decision: in the morning I would tell my Peace Corps trainers to send me back to America. Except—I was in Kenya. Kenya sits on the equator. From there, I could see stars that were always below the horizon in northern climes like Michigan. In fact, I had always wanted to see the Southern Cross. If I were to leave tomorrow, then tonight would be my last chance to see it. I quickly looked up in my book of constellations when the Southern Cross would be visible. If I got up at four in the morning, I should be able to see it just rising far to the south. So I set my alarm and prayed for clear skies.

At that hour I got up, pulled on my clothes, went outside, and saw the famous southern constellations: the Centaur, the Wolf, and, on the horizon, the Southern Cross! It was delightful. Satisfied that I could go home in peace, I turned to head back to my bed.

And there in the sky to the north, gloriously in view as I walked to my room, were all the constellations I remembered from my childhood. The Summer Triangle, rising in the east. The winter constellations, Taurus and Orion and Gemini, setting in the west. The Lion, as bold as any in the Kenyan game parks, prancing overhead. And the Big Dipper to the north, with its pointer stars indicating the spot on the horizon where Polaris could be found.

Surrounded by the stars I had grown up with, I suddenly felt at home. No matter where on Earth I traveled, I would be surrounded by familiar stars. I knew where I was. Africa, Boston, Michigan—it was all the same universe.

What a wonderful universe I was living in! *Let's go!*

And so I stayed to teach for two delightful years in Kenya. When I finally returned from Africa, I got a teaching job at a small school in Pennsylvania, Lafayette College, and started trying to fill my students with the same joy and enthusiasm I had discovered for myself in Africa. Although I loved being the young professor, something was missing. When I attended the wedding of Peace Corps friends (whose schools in Africa I had once visited with my little telescope), I was delighted for them, but, oddly, I did not envy them.

What to do now? Once again, I turned to prayer. Once again, God surprised me. I heard a suggestion that I had never considered, never even thought of before, although in retrospect it seemed obvious. I knew I did not have a priest's vocation, and I'm pretty clueless when it comes to helping people with their personal problems. But the Jesuits also had brothers as well as priests. Brothers aren't ordained, so they aren't expected to lead public prayers or necessarily do pastoral work. Often, they are janitors, cooks, bus drivers. Others are skilled artisans, painters, and builders. Maybe they could also be scientists?

As a brother, I could live in a community alongside the Jesuit priests, with the Catholic sacraments that I so loved built into my daily life. I could teach. I might even do scientific research, although I assumed that my research days were behind me.

Was I crazy to think these things?

The reaction of my family, my friends, even some of the women I had once dated, was uniformly the same: "Of course. That's where you belong. We could have told you that years ago." When I entered the Jesuit order, I discovered that it was indeed what had been lacking. I had been a happy enough person before, but now I had become *content*. I had had a good run at being a scientist. But that life, I felt, was behind me. I would be a teacher and find God in my students, not in rocks and planets and stars.

So of course, after my novitiate and philosophy studies, once again God confounded me. I was handed my first assignment as a Jesuit brother—my only assignment thus far. I was ordered back to doing research in astronomy, using the telescopes and the meteorites at the Vatican Observatory.

No matter where on Earth
I traveled, I would be surrounded
by familiar stars. I knew where
I was. Africa, Boston, Michigan—
it was all the same universe.

Open Cluster NGC 609 as imaged by the Vatican Advanced Technology Telescope.
This cluster of stars is located about 1,300 light-years away from Earth.
Image by Br. Jonathan Stott, SJ.

CHAPTER THREE
THE STARS IN THE SPIRITUAL EXERCISES

MY JESUIT PATH TO THE STARS incorporated examples of classic Ignatian spirituality a lot more than I realized at the time. There are many ways to encounter God, but the core of what makes the Jesuit way special can be found in the *The Spiritual Exercises of St. Ignatius*. Doing these Exercises is a key part of the training of every novice Jesuit. And how those Exercises came to be is like a story out of a fantasy novel.

On May 20, 1521—a little over five hundred years ago—a glory-mad young soldier named Íñigo de Loyola was part of a squad defending a fortress in the north of Spain against attacking French troops. Hopelessly outnumbered, the officers in charge of the defense quickly decided to surrender the fort and avoid pointless bloodshed. But Íñigo was so filled with a desire for glory that he convinced the troops to fight even in the face of certain defeat.

Almost as soon as the battle began, a cannonball smashed into Íñigo's leg, taking him out of commission. The fort immediately surrendered, and Íñigo was sent home.

Lying in bed for months waiting for his leg to knit, Íñigo was forced to spend time thinking through who he was and who he wanted to be. It was the start of a ten-year spiritual journey that eventually led him, by then in his forties, to the study of theology at the University of Paris. There he adopted the name Ignatius and began sharing with his fellow students the spiritual lessons he had learned.

What he shared with them, the heart of Ignatius's spirituality, can be found today in the book *The Spiritual Exercises of St. Ignatius*. The Spiritual Exercises involve a series of prayers and reflections organized to be given during a thirty-day period called a retreat. During that time, retreatants live separated from their daily life, usually at a pastoral

> Three themes underlie the Exercises: awareness, Scripture, and imagination.

retreat house, and maintain silence except for a daily consultation with a trained spiritual director. The common theme of the Exercises is to come to recognize and identify God's presence in specific times and places in human history and in one's own life.

Three themes underlie the Exercises: awareness, Scripture, and imagination. We are guided first and foremost to a deeper awareness of God in our world. Then we are drawn to see how God is revealed to us in Scripture. This insight prepares us to engage in our new awareness by exercising our imagination to make that presence real to us.

The "Principle and Foundation" opens *The Spiritual Exercises of St. Ignatius*. It calls us to pay attention to our place in creation: why we are created, why the world has been created, and how we relate to creation. Ignatius tells us the following:

1. We are created to praise, reverence, and serve God.

2. The things of this world have been created to help us accomplish this calling.

3. Therefore, we are to use only those things that lead us to this goal, being indifferent to the things that lead us away from this goal.

4. We are not to desire things like health, riches, honor, and longevity of life, but rather are to desire only that which helps us become what God created us to be.

Notice what is common to each of these principles: "created things." We are created; we are "things." Everything in this universe has been created. These created things can lead us to God.

The first week of the Exercises is a time to become aware of our sinfulness. But it is not a period of self-flagellation. Rather, it is only when we recognize our own sinfulness—how much we have tried to mess up our relationship with God—that we can realize just how deeply God loves us anyway. He does so freely and because he wants to, not because of some false sense that somehow we could ever do anything to "earn"—or lose—that love.

The second week of the Exercises opens with what is sometimes called the "Trinitarian Gaze." We imagine the divine Trinity looking at this sinful world and pondering what to do about it: "The Three Divine Persons looked at all the plain of the world, full of men, and . . . determined in Their Eternity that the Second Person shall become man to save the human race." And then we are encouraged to use our imagination, that same imagination that got me hooked on science fiction and from there hooked on science itself. As Ignatius writes in the *Exercises,* "See the place: the great capacity and circuit of the world."

> We imagine the divine Trinity looking at this sinful world and pondering what to do about it.

In effect, we imagine Earth as seen from space. Seeing the world from that perspective must have been revolutionary back in the mid-sixteenth century. But, of course, in our era it's a view that we've come to identify with astronauts in orbit.

The space shuttle Atlantis, August 2009; NASA image.

Moving from darkness into sunlight, notice in the view from the Space Shuttle the atmosphere of our Earth lit up by the Sun, with Earth in shadow. The sunlight passing through the air produces a curved blue band of light. This thin band is the locus of our earthly life. It is the band that we are busy poisoning with our human sins of abuse and pollution. And it is the place where the history of our salvation has taken place, culminating in the Incarnation.

The use of the imagination is key to the Spiritual Exercises. Recall what Scripture tells us about creation: the Genesis story of seven days, the story of Adam and Eve in the garden, the story of Job, the story of the mother in Maccabees who tells her sons to trust their God who had made the universe from nothing. Notice also how all these profound insights into the nature of reality and

> The use of the imagination is key to the Spiritual Exercises.

its meaning are told as stories. Then our awareness of creation leads us to reflect on our moments of meeting God, and our stories connect each moment with how and when and where we were when we became aware.

And once we spell out what it is that we are expecting to see, we can discover where our imagination may lead us—or may lead us astray.

The central weeks of the Spiritual Exercises take moments in the history of how God has interacted with humanity, for instance, as described in Scripture, and how God has interacted in our own lives. We are invited to use our imagination to re-create those scenes in our own minds, in effect watching from the outside as we relive them.

But there's a final week to the Exercises, one that too often a typical retreatant never gets to. In the Fourth Week (which usually lasts only a couple of days, not a full week) we sit back and contemplate the fact that in the life of Jesus, after all the illnesses cured and all the opposition and jealousy encountered, and after the Passion itself, there came the Resurrection. Too often Christianity is attacked for promising some unlikely future in which everything will be just fine if we wait long enough. But that misses the whole point. Christianity actually teaches that this is not something that we must wait for. *It's something that already happened.*

Of course, we all need to grow into living the Resurrection—easier said than done, to be honest, because doing so means living without fear.

What does Jesus constantly command in all the stories after the Resurrection? "Do not be afraid." Don't be afraid of poor people; don't be afraid of freshman physics. Don't be afraid of death; don't be afraid to be alive. Don't limit yourself with the lies you tell yourself: *I can't do that, I can't live in a foreign country, I can't solve physics problems, I can't be pastoral to others.* What's stopping us from doing those things except our own fears of failure and inadequacy? Of course we're inadequate! But that is precisely what forces us to make room for God to enter in and help out.

If we knew it all, there'd be nothing left to learn. If we could do it all, there'd be nothing left to do. If it weren't hard, it wouldn't be an accomplishment; it wouldn't be any fun.

Meanwhile the stars are out there above us, beautiful always.

Found in the Fourth Week of the Spiritual Exercises is a famous prayer, "Take and Receive," considered by some to be the most difficult prayer of St. Ignatius. It goes:

> Take, Lord, and receive all my liberty, my memory, my understanding, and my entire will, all that I have and call my own. You have given all to me. To you, Lord, I return it. Dispose of it wholly according to your will. Give me only your love and your grace, that is enough for me.

Recall from Matthew's Gospel (19:13–29) the famous story of the rich young man who wants to know the answer to the question, "What good thing must I do to get eternal life?"

He thinks he knows the answer; it is to follow the Law, like he has been doing. Instead, he is shocked by the answer he hears. It's not enough to follow the commandments, to just do what we're expected to do. Rather, he has to be ready to give up all his other possessions.

This teaching is what is echoed in Ignatius's prayer. Be ready to give up everything. After all, none of it is ours to begin with.

For those of us entering religious life, it might seem that the bargain of the rich young man is not so hard. Giving up possessions is relatively easy, because the unspoken assumption is that we'll be fed and clothed, somehow, by our religious community. But that's not what Ignatius is asking here.

Give up our *understanding*? Isn't that contrary to what being a scientist is all about?

> Be ready to give up everything. After all, none of it is ours to begin with.

Not really. In fact, it's an essential element to doing science. You have to be willing to give up being satisfied with the way you used to understand something in order to understand it in a new and deeper way.

Ignatius's *Spiritual Exercises* calls us to be rooted in this universe while never forgetting that we are directed and motivated by something that is beyond this universe, something that is supernatural and outside the space-time continuum. This is *incarnational spirituality*: allowing the natural to lead us to the supernatural. That is why we study, and teach, the physical universe in all its forms and manifestations.

We are invited to look down on this creation with the same perspective as that of the Trinity, looking down at all the plain of the world, and to look up to the rest of the universe as well. Why? Because when we gaze upon the cosmos, we gaze upon the place where we encounter God.

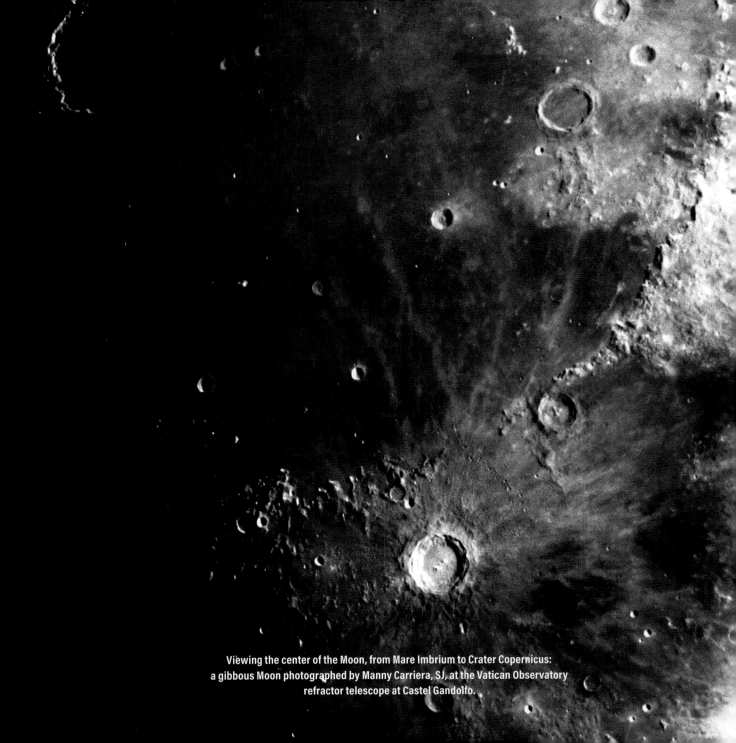

Viewing the center of the Moon, from Mare Imbrium to Crater Copernicus:
a gibbous Moon photographed by Manny Carriera, SJ, at the Vatican Observatory
refractor telescope at Castel Gandolfo.

CHAPTER FOUR
MARVELOUS THINGS WITHOUT NUMBER

IN THE BOOK OF JOB, we are told of an incident when God challenges Job to view his creation, and Job feels too humbled to go any further. Job says, "How can a mortal be just before God . . . who alone stretched out the heavens . . . ; who made the Bear and Orion, the Pleiades and the chambers of the south; who does great things beyond understanding, and marvelous things without number?" (9:2, 8–10).

When was the last time you contemplated the marvelous things without number that God has done in this universe? Do you know where to find the Bear and Orion, the Pleiades and the chambers of the south? Those are constellations, after all—things in the sky that in principle anyone can see for themselves.

In *The Discarded Image,* based on his literature lectures at Oxford and Cambridge, C. S. Lewis insisted that to understand how the universe looked to medieval and Renaissance authors, we "must go out on a starry night and walk about for half an hour trying to see the skies in terms of the old cosmology." Here are some exercises, specific things you can do to make yourself more aware of the stars overhead.

Learn to see the sky. Go outside every night at about the same time and keep a diary of what you see. Find a good place to look at the stars. You must go outside; you cannot really see the sky through a window (as all prisoners know), so why imprison yourself indoors at night? The place where you observe should be dark, away from city lights; it should be open, away from trees and tall buildings; it should be easy to get to.

Is the Moon visible? Where is it located? Is it in a different place, or a different shape, from last night? If you can't see it in the nighttime sky, look for it in the daytime!

Where is the brightest star you see in the sky? Describe its position in terms of cardinal direction (north, south, east, and west) and approximate height above the horizon. Using your fist, at arm's length, as a standard measure, figure out how many fists above the horizon the star is.

Can you find the Big Dipper or Cassiopeia—five stars in the shape of a big W? Throughout the year, one or the other is always easily visible in the northern part of the sky. (If you live too far south of the equator to see them, look for the Southern Cross.) Where are they located? How big across are they in terms of your standard measure of fist at arm's length?

Can you see any stars that have a color other than silvery-white? If so, where are they located? Can you find any clusters of stars, all bunched close to each other like a little cloud of light—until you look more closely?

If you have a clear horizon to the east, find a bright star (or the Moon) near the horizon and watch it rise. Notice the stars that are visible when you start gazing and where they are when you finally go indoors, half an hour later at least. If you have a clear horizon to the west, watch some star (or the Moon) as it sets.

Imagine you are an ancient Chaldean shepherd out at night with your sheep, gazing up at the stars. What do you think you would make of them? Imagine you are a medieval scholar, living at a time when the commonly accepted view of the universe describes the sky as a fixed orb of stars overhead and the planets as attached to transparent rotating spheres between those stars and us. Do you feel a sense of up and down? A sense of absolute smallness compared to the universe?

This is an act of prayer we're talking about, so treat it as such. Do it regularly. Do it in silence. Do it alone. Do it (at first) without a book or binoculars or any other sort of extra things. Learn to see stars not only with the naked eye, without a telescope, but also without any artificial expectations. Do it often enough that you recognize the same stars every night.

Find the rhythm of the stars. Notice how the Moon's position changes over the course of a month; notice how the Sun's position changes over the year; notice how different stars arrive and depart over the course of a year. Notice how stars' positions shift as you go north or south. Notice how planets' locations change over many months and many years.

You have to see in order to understand.

Why should we do this? What will it teach us? It teaches us to recognize regularity and change: we must know what's regular before we can recognize what is unusual. It teaches us to recognize pattern and surprise: we must know what the patterns are before we can recognize what is outside of the pattern.

You have to look in order to see. You have to see in order to understand. Seeing is more than just looking—it's paying attention, which is hard. But it's also spiritual. Learning to pay attention is learning to pray.

The sky is like a friend, a lover. You may never completely understand it, but there is great joy to be found simply in spending time with it, getting to know its moods and its smiles.

Were you able to find a really good dark site? After spending about twenty minutes there, was it as dark as it seemed to be when you first went outside? Or did the night sky begin to disappoint?

Light pollution is the culprit. Dark Sky International (darksky.org) has studied light pollution in detail and classified its many insidious forms.

Glare is the excessive brightness from nearby sources, like a streetlamp or a neighbor's outdoor light located in just the wrong place. It's local, and it bothers your eyes. One particularly glaring problem is light from poorly shielded outdoor streetlamps. If you were to go to your friend's house in the evening and discover that their living room was lit by bare lightbulbs shining directly in your eye, you'd think they were nuts. But that's how we light our streets. If you can see the source of the light, then the light is shining in your eyes instead of down on the ground where you are trying to drive or walk. Worse, that kind of lighting decreases your vision by blinding you with glare and turning shades of gray into stark blacks or whites. This limits our ability to see potential dangers at night. Aging eyes are especially affected by this.

Skyglow is more universal. It is the general brightening of the night sky that happens near inhabited areas. In fact, its effects can be felt hundreds of miles away from the city. You may not even notice that it is there, but it's everywhere. Recall the famous opening of a novel: "It was a dark and stormy night." Well, that's not true anymore. Stormy nights in light-polluted areas are no longer dark; clouds reflect the light of nearby cities, bathing everything in a sickly gray glow. In fact, about 80 percent of the world's population lives under skyglow—and in the United States and Europe, 99 percent of the public no longer experiences a natural night!

Light trespass is when light falls where it is not intended or needed. While skyglow is the more general case (streetlamps that light up clouds instead of the street), locally you can also have spotlights on a building that spill over onto other nearby buildings—and into your bedroom window if you are unlucky. Good luck sleeping.

Clutter is when there are bright, confusing, and excessive groupings of light sources. That's common on city streets at night. It can be hard to make out traffic lights or street signs against all the other blinking lights that grab our attention.

The effects of light pollution go beyond making it hard to see the stars. Every bit of light that hits a cloud instead of a street is wasted: I don't need to see where to walk on a cloud. The Dark Sky folks estimate that the electricity consumed by light pollution runs to at least $3 billion wasted per year.

Light pollution disrupts ecosystems and wildlife. Both predators and prey can become confused. Migrating birds can fly themselves to exhaustion circling tall lit buildings that they mistake for moonlight. Newly hatched turtles have been seen scurrying at night toward overly lit resort hotels rather than where they belong, in the ocean.

And light pollution affects us personally. Our sleep rhythms are disrupted (especially by blue LED light), which makes it harder to get to sleep and stay asleep. In the end, the irony is that we put up city lights to make ourselves feel safer, but in fact they often do exactly the opposite. They distract us with clutter and cast a sharp, dark shadow over everything.

For most of human history, nightfall meant the absence of light, a daily shift in what we could and could not do. Darkness meant retreating within our homes, sitting together around a fire, waiting for dawn. Thus, light and dark become images of our moral journey.

Nowadays it's rare that any of us ever gets to see the sky as it really is. There's a famous story that after the 1994 earthquake in Los Angeles, hundreds of frightened people called the Griffith Planetarium wanting to know why the quake made the sky so scary. They'd gone out of their houses before dawn when all the city lights were off, and for the first time they had seen what the sky actually looks like. It was terrifying! Isaac Asimov wrote a story called "Nightfall" about a similar experience, but what he set on a planet with multiple suns in fact has already happened to us: we've blinded ourselves with our own artificial suns.

I'm an astronomer. You would think I would be used to seeing the dark sky. But to see the stars, astronomers now have to make extraordinary efforts to get away from city lights. Telescopes are placed on mountaintops far from cities, and cities with lots of astronomers, like Tucson, have laws—not perfect but better than nothing—to try to limit the spread of skyglow. One of the best views of the sky I was ever able to see was at the Okie-Tex Star Party, held in the far-western panhandle of Oklahoma. I could see my shadow from the light of the Milky Way.

> Nowadays it's rare that any of us ever gets to see the sky as it really is.

Once there was a time when anyone, anywhere, could have seen that.

When I was with the Peace Corps visiting friends in rural Africa, the day was counted with twelve hours starting at sunrise, not midnight. (Living on the equator, sunrise happens at the same time of day, year-round.) Where most of my Peace Corps friends lived, away from the city, sleeping and waking were controlled by the rhythms of the sky, with occasional help from candles and kerosene lamps. And for people who subsisted on the food they grew, kerosene cost money, which was always in short supply.

Like the "wise virgins" of the Gospel story, oil for your lamp was something you had to buy in advance and parcel out sparingly.

With artificial light has come a number of profound cultural changes that we barely notice. Artificial light has changed our reading habits, our working habits, our sense of safety on the streets at night. And most importantly, it has changed how we understand images of light and dark in Scripture and literature.

"Let there be light." But it's all light now, and we can't get away from it.

"The people who lived in the darkness have seen a great light." Great lights are now boring.

"I am the Light of the world." So what?

All those statements carried a whole lot more punch when light was not something we could command with the flick of a switch.

The Easter midnight liturgy starts with a dark church and the entrance of the Paschal candle; several of the readings work off the image of a new light in the world. In the homily of Pope Benedict on Holy Saturday, the one I mentioned previously, Benedict commented:

> What the Church hears on Easter night is above all the first element of the creation account: "God said, 'let there be light!'" . . .
>
> What is the creation account saying here? Light makes life possible. It makes encounter possible. It makes communication possible. It makes knowledge, access to reality and to truth, possible. And insofar as it makes knowledge possible, it makes freedom and progress possible. . . . The darkness that poses a real threat to mankind, after all, is the fact that [we] can see and investigate tangible material things, but cannot see where the world is going or whence it comes, where our own life is going, what is good and what is evil. If God and

moral values, the difference between good and evil, remain in darkness, then all other "lights," that put such incredible technical feats within our reach, are not only progress but also dangers that put us and the world at risk. Today we can illuminate our cities so brightly that the stars of the sky are no longer visible. Is this not an image of the problems caused by our version of enlightenment?

Pope Benedict sees light pollution as an analogy of sin. To me, it is more than an analogy; it *is* a sin. Putting up a glaring streetlight robs us of the dark starlit sky given by God to inspire and comfort us all.

In spite of light pollution, with a little effort we can still travel to dark-sky sites where the Milky Way can be seen. What's more, nowadays we can purchase a small amateur telescope at a surprisingly low cost that gives us a view, with our own eyes, of some amazing sights in our galaxy, even in some cases in light-polluted skies.

When I returned from Africa, I asked my friend Dan Davis (who had helped me buy a telescope to take to Kenya) to help me write a book about what a small telescope can see in the sky. We called it *Turn Left at Orion,* and since its publication in 1989, it has become quite popular. In its opening chapter, we present a brief introduction to the different kinds of things that an amateur's small telescope can actually see. These different types of astronomical objects are sights worth looking for, even with just binoculars, so it's good to learn what they are and what we call them. They are also the stuff of which our universe is made. Shown on the following pages are sketches from *Turn Left at Orion* of these objects as seen in a small telescope, side-by-side with color images of the same objects illuminated with the Vatican Observatory's six-foot-wide Vatican Advanced Technology Telescope mirror.

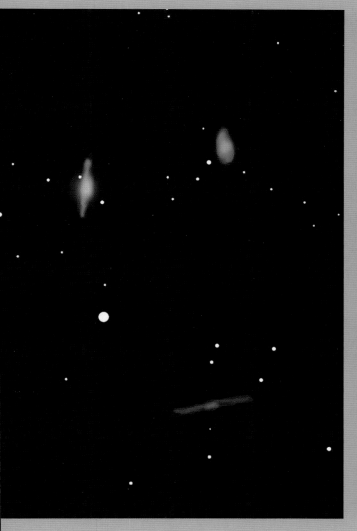

Three galaxies as they look together in the same field of view of a small telescope. They lie within the constellation Leo. The smudge to the upper left in this image is designated M66. All these small telescope images are by Dan Davis and adapted from pictures found in *Turn Left at Orion*.

The galaxy M66, about twenty million light-years away from us, as seen by the Vatican Advanced Technology Telescope.

The globular cluster M56 as seen through a small telescope. Located in the constellation Lyra, this ball of 100,000 stars is ten light years in diameter and lies about 40,000 light years away from us.

This is the globular cluster M56 as imaged in the Vatican Advanced Technology Telescope.

The fundamental building blocks of the universe are *galaxies.* These are billions of stars in immense assemblages, a thousand light-years across, millions of light-years distant from us. As we note in *Turn Left,* "It is astonishing to realize that the little smudge of light you see in the telescope is actually another 'island universe' so far away that the light we see from any of the galaxies that we talk about (except the Magellanic Clouds) left it before human beings walked the Earth." (See images on page 57.)

The stars within our own Milky Way galaxy can be found in various configurations, too. First are *globular clusters*—spherical groupings of about half a million stars, held together forever by their mutual gravitational pull in a densely packed swarm of stars. (See images on the previous page.) The stars in globular clusters have turned out to be very old. They apparently are the first things made when a galaxy forms, and they are generally found around the center of the galaxy. Globular clusters are glorious to look at in a little telescope—they appear as eerie spheres of white haze that can just barely be resolved into tiny points of individual stars if the conditions (and telescope) are just right.

But, of course, the bulk of the stars in any galaxy are not those in clusters; they are spread out (usually) into a disk or an elliptical cloud of stars, which forms their galaxy.

Stars shine by fusing their hydrogen atoms into helium, emitting the energy that results as starlight. Eventually stars run out of hydrogen to fuse. The last gasp of energy emitted as they burn out and collapse puffs out a cloud of hydrogen and oxygen gas that sometimes takes on the shape of a disk, sometimes a ring—the edge of the cloud is denser than its insides, so we see its shell as a ring surrounding the dying star. We call these *planetary nebulae.* (See images on page 61.) They have nothing to do with planets, but they look like small disks and can be easily mistaken for planets if you don't know any better. The gas glows both red, from the hydrogen, and green, from the oxygen. The red shows up better in color images, but the green is easier for the human eye to see.

All that puffed-out gas from the various dying stars eventually collects itself into what's called a *diffuse nebula*. (See images on page 62.) Diffuse nebulae can be eerie, swirly clouds of red and green light, and they are glorious to see in a small telescope if you are at a good dark site. The more gas, the more gravity; the more gravity, the more the gas can fall back together into lumps that become clusters of newly formed, second-generation stars. Thus, sometimes a cluster of pinpoints is visible, each one a young star being formed within that gaseous womb. Once the gas is completely sucked into the stars, an *open cluster* of young stars is left. (See images on page 63.) The easiest open cluster to find is the Pleiades, one of the sights mentioned in the book of Job.

As Dan and I said in *Turn Left,* "Viewing an open cluster can be like looking at a handful of delicate, twinkling jewels. Sometimes they are set against a background of hazy light from the unresolved members of the cluster. On a good dark night, this effect can be breathtaking." But even if the sky is not particularly dark, they're relatively easy to spot and observe in a small telescope or binoculars.

The sky contains marvelous things without number. Our city lights have done their best to limit those countless things to the Sun, the Moon, and a few bright stars. How can we find God in all things if we willfully blind ourselves to some of the best of those things?

And yet despite our worst efforts, the sky is still worth looking at. Right now, as you are reading these words, from where you are sitting, is it cloudy or clear? Is the Moon visible? When did the Sun rise today, and when will it set? (See page 64 for a photo of a sunrise with the Moon still visible in the sky.) Are you aware of these things, or are you trapped in a mental room without windows and doors?

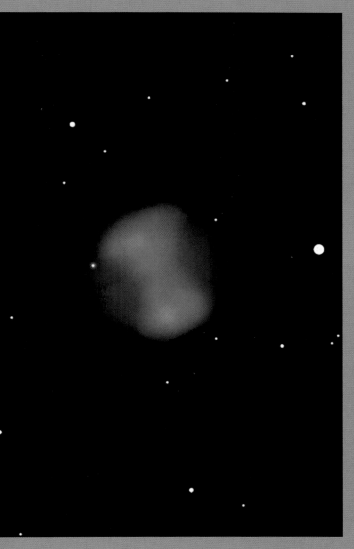

The Dumbbell Nebula, designated M27, is a bright planetary nebula found inside the Summer Triangle of the bright stars Deneb, Vega, and Altair. It's about 2 light-years wide and lies 1,200 light-years away from us. This is what it looks like in an amateur telescope.

This is M27, the Dumbbell Nebula, as imaged at the Vatican Advanced Technology Telescope.

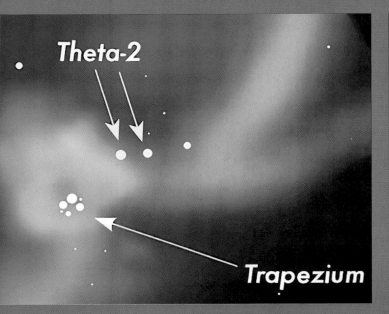

Theta-2

Trapezium

Perhaps the most famous diffuse nebula is the Orion Nebula, about 1,300 light-years away in the "sword" of Orion, the bright winter constellation. Near the center of this star nursery is a cluster of young stars centered on four bright stars called the Trapezium; nearby, the stars identified as Theta-2 are in orbit around each other. This drawing shows what you can see with a small telescope.

This is the Orion Nebula as imaged with the Vatican Advanced Technology Telescope.

This open cluster, M103, is located about 8,500 light years away from us. It is one of many open clusters found in the Milky Way near the constellation Cassiopeia. This is the view with a small telescope.

The Vatican Advanced Technology Telescope view of the open cluster M103 can distinguish some of the stars as orange or red; these are more massive stars that have already evolved into their "red giant" phase. We can use them to estimate the age of the cluster—the more red stars, the older the cluster.

A thin, waning crescent Moon at sunrise, near the Jesuit Community of the Vatican Observatory in Tucson, Arizona.

CHAPTER FIVE
STAR RELICS

ONE OF THE COURSES THAT I TOOK as a student at MIT shaped my career in more ways than I could ever have imagined. It was taught by Dr. John Lewis, who later directed my research projects at MIT. The topic of the course was meteorites.

As I mentioned in a previous chapter, I had been surprised when I first arrived at MIT and discovered that by joining the Department of Earth and Planetary Sciences, I had actually signed up to study geology. I worried that rocks might be boring. Fortunately, my first classes in geology reassured me that minerals were actually grand for seeing basic physics and chemistry in action. But even that didn't prepare me for the joy of hearing that I could hold pieces of outer space in my hands.

Meteorites have been falling to Earth since the time the planet was formed. In a true sense, Earth itself is a large ball of meteorites that has melted and differentiated: dense iron and nickel at its core, basic rock over that, and a coating of snow and ice and air and carbon-rich soil (and less-dense rocks like feldspars and granites). All these ingredients are found in meteorites in their raw state, undifferentiated. (See page 66 for a photo of a meteorite.) Some meteorites have feldspars and flecks of iron intimately mixed together; others are permeated with seams of clay that are rich in water and carbon.

I learned how the decay of radioactive elements in these rocks can be used as a clock to date when the meteorites first formed and to measure when they were last disturbed by any kind of chemical or physical event. From this kind of dating, we know that most meteoritic rocks date back to the earliest times, when the solar system itself was being formed, four and a half billion years ago. This makes them valuable witnesses for when we try to figure out what was going on during the era when the planets were being made.

To me, meteorites also hold another significance. They are actual bits of outer space! The planets that we talk about in our meetings, the planets that we see overhead in the sky—they are real places, places made of real stuff, of things we can touch and taste and smell!

Yes, smell! A meteorite recovered in January 2000 from the frozen Tagish Lake in northern Canada gave off a distinct odor when it was warmed up to room temperature in the lab. (My colleagues in Alberta eventually built a refrigerated lab to work on the sample.) And yes, taste! I know a meteorite dealer who likes to taste the dust that gathers when he cuts bigger meteorites into smaller pieces to sell to collectors. Why? Just because he can, because he is also amazed that these really are samples from billions of pieces of rock that have spent billions of years in orbit around the Sun, billions of kilometers from Earth.

A slice of the meteorite Esquel, from the Vatican collection. This is an example of a pallasite meteorite featuring translucent olivine crystals (backlit in this image) embedded in iron-nickel. This slice is about ten inches wide and weighs almost half a pound.

Does that remind you of anything in the world of faith? Well (except for tasting), there's a reliquary off to the side of St. Peter's in Rome, a room full of the bits of bone and cloth and other things associated with saints. The saints we honor with incense and feast days were humans just like us, walking the same paths, breathing the same air. You can actually touch things that they once touched. And most unnerving of all, Jesus himself, the Incarnation of the Creator, also walked and breathed in the same universe where we walk and breathe.

> Holy relics and meteorites remind us, though, that the past is actually in the same place as the present.

The Kenyan philosopher John Mbiti, in his seminal book *African Religions and Philosophy,* describes how people in East Africa sometimes think of time as two places, the past and the present, without our western idea of an unbroken line connecting the two. And I can certainly understand that point of view. My beloved science fiction stories often feature the idea of "travel" to different times, as if the past were a geographical place. Indeed, when a familiar song brings me back to my university days, my sense of nostalgia is so strong that it seems I should somehow be able to go back there physically, as if it were a place that I could visit. Holy relics and meteorites remind us, though, that the past is actually in the same place as the present.

Copernicus and Newton, the founders of modern science, gave us the revolutionary idea that stars and planets are made out of the same stuff as us here on Earth, subject to the same laws of physics and chemistry. We can study meteorites with exactly the same tools we use to study Earth rocks. We can ask the same questions about them. We can use the same theories and measurements to get to know them better. That's why I was working in the same department at MIT that also studied Earth geology. Meteorites may be rocks from space, but they are still rocks.

Back in the early 1900s, a French gentleman and scientist, Adrien-Charles, Marquis de Mauroy, amassed one of the largest private collections of meteorites in the world. (He did it the old-fashioned way—he bought them from dealers.) (See page 73 for a photo of the Vatican Meteorite Collection.) A devoutly religious man, he offered to donate a selection of meteorites to the Vatican to help start a museum of natural history. According to the columnist "Tiber" in Milan's newspaper *l'Unione,* published on December 20, 1907: "When Cardinal Mocenni was told about de Mauroy's intentions, he drew deeply on his pipe—his inseparable companion at home and in the office—and exclaimed with a sarcastic smile, 'But what is it to us to have more pieces of rock, large or small? Don't we already have them in abundance in Italy, so many as a matter of fact that we could send to France as many as they want?'"

It's easy to laugh at the cardinal. But if we were to try to answer him seriously, we would need to outline how meteorites are *different* from Earth rocks. How can we tell a meteorite from an Earth rock? That's a very interesting sort of question to ask.

Would you have difficulty in describing how a rock is different from a chicken? Or a book? Or the ocean? In such cases the differences are so obvious that you wouldn't even know where to start. But exactly because the difference is so obvious, the question never comes up. Ironically, it is only when the differences are minor that they become important.

To know what is special, you must know what is normal. In the history of geology there was a great debate beginning in the nineteenth century about how geological features were created. Some insisted that the geology of Earth's surface changed only gradually, over a long period of time: this theory is known as uniformitarianism. Others insisted that change occurred suddenly and violently: catastrophism. At the time, this was seen as a debate between those of a mechanistic point of view who put their faith in "processes" and natural laws, and those who looked for God to send earthquakes and floods.

But in the geology of other planets, where the scars of events that occurred billions of years ago can remain visible, untouched by the wind and rain that erodes their traces on Earth, we see that the most common form of geological feature is the impact crater. Can you imagine anything more catastrophic than the sudden collision of a large meteorite? Meteorites can hit the surface of a planet at a speed of more than a hundred thousand miles per hour. The energy that is released in such a collision can be greater than the energy of the largest nuclear bomb.

And while we rarely see such collisions in a human lifetime, on the scale of the age of the solar system they occur all the time. In other words, both the uniformitarians and the catastrophists were correct. Catastrophes do happen—all the time, uniformly.

Are catastrophes like meteorite impacts acts of nature or acts of God? Famously, one giant impact sixty-five million years ago is thought to have driven the dinosaurs to extinction. Was that an accident of nature or an act of God?

But isn't it the case that every act of nature is also an act of God? After all, God set up the universe and created the laws of physics that we identify as "natural." But could you argue then that every act of God is an act of nature?

What does this tell us about miracles? Nowadays, we often think that a miracle is a moment when God suspends the laws of science to act directly in the universe. But think about this: our Scriptures talked about miracles long before anyone had even thought of a law of science. How can the nature of a miracle be tied up with some notion of a scientific law when we had miracles before we had laws?

To me, miracles have nothing to do with scientific laws. Rather, a miracle is any remarkable sign that serves to direct our attention to God. That sign does not have to be a violation of a natural law; it just has to be something that will get our attention, something unusual, something different from what is normal. After all, the impact of a meteorite falling on your head would certainly get your attention, but it would not be in violation of any natural law.

Just because winning the lottery is rare doesn't mean that it can't ever happen. Somebody has to win the lottery, even if it is not likely to be you or me. And winning the lottery isn't necessarily an act of God. The other essential part of a miracle is that it directs our attention to God. Too often, though, suddenly getting rich can have the opposite effect. We might even call that an antimiracle.

We also have to remember that a law of nature is simply a human approximation of how nature itself actually behaves. The more we learn about nature, the more we need to refine how we describe these laws. Newton came up with laws of physics that do a really good job of describing many things in nature; they are good enough that we can use them to design airplanes that fly and bridges that don't fall down. But since the beginning of the twentieth century, we have discovered all sorts of extreme cases in which Newton's laws don't work, whether subatomic particles or black holes.

Modern physics has developed two new sets of laws of nature to handle what we see in these extreme cases: general relativity and quantum mechanics. And one of the exciting (or embarrassing) features of modern physics is that, so far, nobody has been able to come up with a way to make the laws of general relativity agree with the laws of quantum mechanics! This does not mean that we won't eventually come up with some way to understand how they mesh; it's just a reminder that our understanding of what we call the laws of nature is not complete. And indeed, it never will be complete. The more we learn, the more we discover is yet to be learned.

Even in the realm of planets, which (mostly) is not so extreme that Newton's laws fail us, we see that our understanding of what is normal and what is unusual can be confounded.

In that course I took in my undergraduate years about planetary physics and chemistry, one of the wonderful things I learned back then, some fifty years ago, was how we could set limits on how planets were formed just by looking at the regular trends we see in our own solar system. For instance, we can see that planets do not travel every which way when they go around the Sun; rather, they all travel in the same direction, which is also the direction that the Sun spins, and they all lie together on pretty much the same plane. This strongly suggests that all the planets and the Sun were formed together, at the same time, perhaps out of the same spinning disk of gas and dust. In the fifty years since I took that class, our telescopes have become powerful enough that we can actually see disks of gas and dust that look just like we predicted they should look.

> The more we learn, the more we discover is yet to be learned.

Another thing we knew is that the planets in our solar system are neatly sorted, with small rocky planets lying closer to the hot Sun while the large gaseous planets lie farther away from the Sun. Using the same chemical equilibrium calculations that let us understand the gas in the atmosphere of Venus, we worked out the chemical equilibrium in the gas cloud that formed all the planets in our solar system. And this told us that within a certain distance from the Sun, only rocky materials would exist as solid bits of dust, which would then form rocky planets. But beyond that distance, the gas cloud would be cold enough that all sorts of ice (especially water ice) should form, and that ice would snowball into the cores of planets that could have enough gravity to capture some of the gas in the primordial disk.

This explains two things quite well. It not only shows us why rocky planets and gaseous planets occur in two separate places. It also explains why there were no planets bigger than Earth but smaller than Neptune. Once a planet got big enough to begin to capture gas—bigger than Earth, apparently—all that extra gas would immediately jump them up into a much bigger size range, up with Neptune and Uranus and Saturn and Jupiter.

This powerful, wonderful theory has only one flaw. It is not true. It doesn't work. Since the 1990s we have begun to discover planets in orbit around other stars. We now know of several thousand such planetary systems. And in all of them, we never see that orderly separation we see in our own solar system, with smaller rocky planets near their stars, and larger gaseous planets farther away. In many cases, the large gaseous planets are orbiting closer to their star than small, hot, rocky Mercury orbits our Sun.

What is even more embarrassing is that the vast majority of the planets we have discovered so far around other stars are exactly in that size gap, bigger than Earth but smaller than Neptune, that our old theory had so neatly explained was impossible!

What we thought was normal—our solar system—has turned out to be exceptional. Miraculous, almost.

One of the children's books that I got to read in the MIT science fiction library was T. H. White's fantasy about King Arthur, *The Once and Future King*. The first novel in that series, *The Sword in the Stone*, tells us about when Arthur was a boy. That's when the magician Merlin teaches young Arthur a lesson that astrophysicists often cite: "Everything not forbidden is compulsory."

In other words, in a universe as huge and varied as ours, if you can imagine something happening, then odds are that someplace, at some time, it probably does occur. That is, if it *can* happen, it *did* happen. That's why astrophysicists are so delighted with taking flights of the imagination while thinking about the universe; our imagination gives us things to look for. Or, to put it another way, if you expect to see something but don't ever see it, then maybe indeed it is "forbidden," or impossible. A variant on this principle plays an important role in how we understand certain meteorites that we now think came from Mars.

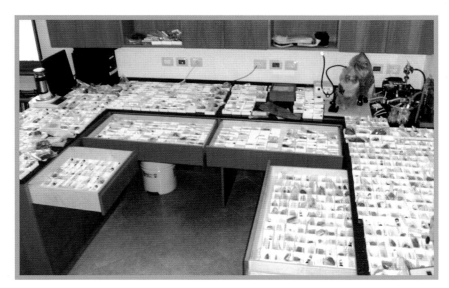

Vatican Meteorite Collection; Br. Robert J. Macke, SJ, took this image of the Vatican's meteorite collection during an inventory in 2018.

We have established that we must know what is normal before we can recognize and understand the unusual. There are certain things we can say are normal about most meteorites; virtually all meteorites differ from the usual rocks you find on Earth in a

The author in the Specola Meteorite Lab.

number of ways. The most obvious way is that most have tiny flecks of metallic iron and nickel scattered in among the grains of rock. Why is that never the case for Earth rocks? For one thing, the denser iron should be separated from the less dense rocky material on any planet, like Earth, that would be hot enough to melt. But even the meteorites themselves that land on Earth today don't keep that metallic iron very long. Earth's air and water quickly cause the metal to rust. Indeed, it is rare for ordinary meteorites to remain intact for more than a hundred years or so (unless they are kept dry in a

museum), since the iron swells up as it turns into rust, and once it expands, it breaks the meteorite apart.

Another difference is that even the very oldest intact Earth rocks are younger than about three billion years old, whereas virtually all meteorites can be dated with radioactive isotopes to be four and a half billion years old, half again as old as the oldest Earth rocks. The reason for this is also easy to explain. The smaller the parent planet of a meteorite (or any other rock), the less that parent planet can hold on to the heat in its interior that might alter or melt the rocks and thus reset their radioactive clocks. Even the youngest rocks from our Moon, the ones that come from the dark lavas we see as the "face" on the Moon sitting atop the older lunar crust, are mostly older than three billion years old. Since we believe that meteorites come from tiny asteroids (except for a few that match Moon rocks in their chemistry and age), it's not surprising that they haven't been able to reset their clocks since they were formed.

Now, while these rules for what is "normal" for a meteorite may be true for the thousands of different meteorites we know about, by the early 1980s we recognized that there were a tiny number of meteorites—about a dozen were known back then—that didn't fit these rules.

First of all, these particular rare meteorites didn't have metallic iron. Their minerals were rich in iron, to be sure, but all that iron was oxidized just like the oxidized iron in Earth rocks. Second, the rocks were all very young, just a few hundred million years old, less than a tenth of the age of the other meteorites.

So how did we even know they were meteorites and not just Earth rocks? Well, for one thing, someone had actually seen many of them fall out of the sky. Also, they were coated with the thin layer of black molten rock that all meteorites have, which comes from how a meteorite surface melts when it plunges into our atmosphere at an enormous speed. And when we carefully examine the elements in these rocks, we find subtle chemical traces that make them stand out from Earth rocks—for example, we can see that the minerals have been exposed to cosmic rays in space for millions of years.

So where did these rocks come from? Because of their oxidized iron they likely came from a place that had an atmosphere. And they would have to come from a place bigger than the Moon, since the rocks themselves were younger than the Moon's rocks. But it wasn't likely to be a place as big as Earth or Venus, since it's hard to see how rocks could be launched from such big planets and into space without being completely destroyed in the process. (We do see Earth material that has been launched into space after big impacts. These are called tektites, and they are no longer rocks but lumps of black glass, melted by the impact that launched them and then quickly frozen as they left Earth's surface. Of course, it's possible that there may be actual intact Earth rocks in the meteorite collection, but if so, they are very rare.)

About the only place that fit the bill was Mars. Granted, Mars is a lot smaller than Earth or Venus, but how could something launch from the surface of Mars? Some parts of Mars are covered with impact craters, so our first guess might be that the energy of the impact somehow shot these rocks off the surface with enough speed to escape Mars's gravity. Could that actually happen?

A while back, the world's leading expert on impact cratering ran the numbers through his models and definitively concluded: No. It was impossible. For one thing, we can compare Mars craters with those we see on the surface of the Moon. Around every big crater on the Moon there are lots of little craters made by the stuff splashed out of the big crater. The bigger the crater, the farther the stuff gets splashed. And we can find craters on the Moon big enough that some of that splashed material stuff should have been ejected fast enough to escape the Moon's gravity. That would account for the meteorites we have whose chemical composition exactly matches the samples brought back by the Apollo astronauts.

But when we scale everything up to account for the size and gravity and atmosphere of Mars, we can conclude that no impact crater on Mars could be big enough to eject an intact meteorite.

Then we started collecting meteorites by searching for them in Antarctica. And one of those meteorites changed everything.

Why Antarctica? It turns out, although meteorites fall everywhere on Earth at the same rate, in most places they get lost among the Earth rocks and eventually rust into dust. But in Antarctica, the black rocks stand out against the white ice. And they don't rust as quickly because it is so cold. Even better, the slow motion of the Antarctic ice cap moves the meteorites downslope toward the sea until the ice encounters a mountain range; at that point, all the meteorites pile up onto a field of blue ice. Since the mid-1970s, annual expeditions have gone to these blue-ice regions of Antarctica to collect meteorites.

The author took part in the Antarctic Search for Meteorites program in 1996. This photo of their camp on the East Antarctic Plateau was taken looking due south at midnight.

The 1979 expedition found one of these putative Martians among the samples they collected. A sharp-eyed geologist examining the specimen noticed that the rock had cracked, violently. The rock had melted in that crack, and then the bit of melt froze itself into glass. And inside that glass were little bubbles. What was in those bubbles?

The researchers carefully carved out a piece of glass with a few bubbles. They put it into the vacuum chamber of a mass spectrometer (a fabulous gizmo full of tubes and pipes and pumps that would be right at home in a science fiction movie). They remelted the glass, and the bubbles burst into the vacuum. They then used the mass spectrometer to measure the chemical composition of the air coming out of the bubbles.

The air they measured exactly matched the atmosphere of Mars. They knew this because the Viking spacecraft had landed on the surface of Mars just a few years earlier and measured its atmosphere. And Viking had found that the Martian atmosphere was quite peculiar and distinctive.

The age of the meteorite matched what you'd expect for Mars. Oxidized iron matched what you'd expect for Mars. And it had bits of Martian atmosphere. The conclusion could not be denied: this rock came from Mars.

What about that theorist? What about all his calculations that said it was impossible for this rock to be launched from Mars? He went back to revise his

This 150g Martian meteorite was donated to the Vatican in 1912, one year after it fell near the Egyptian town of Nakhla.

theories. Even though he didn't understand how it happened, he had to admit: if it *did* happen, then it *can* happen.

A favorite cartoon (by Sidney Harris) among scientists shows a professor at a blackboard working out a mathematical theorem. In the middle of the proof he writes, "Then a miracle occurs . . ." His colleague, looking on, quietly comments: "You should be more explicit here in step two."

No scientist would insist that the arrival of a meteorite from Mars was a miracle. To say so is bad science. And, of course, as we have described, it is bad theology as well. Remember, we have defined a miracle as something remarkable (but not necessarily impossible) that turns our attention to God.

> If meteorites from Mars are not miracles, then what does count as a miracle?

If meteorites from Mars are not miracles, then what does count as a miracle? One obvious example is the Resurrection of Jesus. Without the Resurrection, as St. Paul tells us, the rest of our faith as Christians is in vain. The Resurrection is so startling, so unlikely, that a lot of people have tried to wave it away by saying it was "symbolic," or just some sort of fantasy story that the apostles told to make their religion's founder look good.

But I've read fantasy stories—a lot of them. Many fantasy and science fiction stories treat the same issues that are at the heart of most religions: the tension between good and evil, the true nature of reality, the meaning of existence. But none of them tells stories anything like what you find in the Gospels.

Plenty of them describe some sort of "savior" who comes to redeem their people. A lot of them even wind up reusing elements of the Christian story. For example, in the 1951 movie *The Day the Earth Stood Still,* a flying saucer arrives with an alien who wants to save humankind—he disguises himself as a human and takes the name Mr. Carpenter! But the fascinating thing is that whenever a clever human author tries their hand at imagining how it would work, they never come up with all the odd twists that we find in the Gospels.

In fantasy fiction, the savior or avatar usually arrives with a bang. For example, when the flying saucer arrives in *The Day the Earth Stood Still,* we see it announced by a series of newscasters around the world. But Jesus arrived as a baby, and only the local shepherds and a small group of wise men (and women—don't forget the prophetess Anna) recognized that someone unusual had arrived.

And even when our fictional hero is described as a child, he is sure to display unusual powers. You see this everywhere, from some of the apocryphal gospels to the Harry Potter stories. But the only thing we see Jesus displaying as a child in sacred Scripture itself is an unusual wisdom while speaking among the scribes.

Indeed, the temptations of Jesus at the beginning of his ministry are all standard plot devices of the typical fantasy story. Our fictional hero might turn stones into bread, or lead into gold, or in some other way reverse the flow of entropy for fame and profit. He might throw himself from pinnacles or, like Superman, "leap tall buildings with a single bound." He might achieve dominion over all of Earth, or at the very least leadership of a laboratory that changes the world of science. But Jesus rejects all these plot devices. Indeed, the lesson we should take from this is that when you actually are all-powerful, you don't need to flaunt your power.

Our stories tell us how our heroes become famous. Jesus rejected fame and told his followers not to talk about him (at least while he was alive) but to proclaim the kingdom of God. Our stories tell us how our heroes outwit their opponents; but even when Jesus does answer the attacks against him, cleverly at times, he never uses his advantage to crush his attackers. And when Jesus was given the chance to show them up once and for all, he remained silent before Pilate.

Furthermore, the description of the resurrected Jesus leaves out all the stuff that a fiction writer would want to include: *How does it work? What tricks can I use with this gimmick?* But it includes all the sorts of things that a fiction writer wouldn't bother with: *Why does it take time for his closest friends to recognize him? Why does he have fish for breakfast? Why does he tell the leader of his gang, Peter, that rather than going out and doing great things, he should take up the task usually given to children, and just feed the sheep?*

The stories we read in the Gospels are fundamentally different from the way that any human author could have imagined a savior. And, of course, this insight is not original to me. In *Fear and Trembling*, the philosopher Søren Kierkegaard shows how no human retelling of the story of Abraham's sacrifice of Isaac can match the version we read in Genesis. The patristic theologian Tertullian wrote, "It is certain, because impossible." (The Enlightenment skeptics mocked this as "I believe because it is absurd," missing both the point and the power of the original statement.) I believe what Scripture tells me because no human author would ever have invented what I read there; a purely human author would think the Scripture version was absurd.

The Enlightenment had its literature, as well, but somehow over time the stories of the descendants of the Enlightenment have become stale, humorless, and at times embarrassing. Who can stomach Voltaire's snide arrogance today? And a lot of "golden age" science fiction is still trapped in Enlightenment thinking. I remember loving Asimov's *Foundation* science fiction books when I was a child, but when I tried rereading them recently with an adult's sensibility I was horribly disappointed. They did not age well. By contrast, the stories of Scripture have never lost their ability to startle us. They have never lost their hold on our imaginations.

A skeptic may choose to believe that Jesus never existed or that Isaac and Abraham never existed. But it is hard to argue that the stories about them don't exist! They are miracles: remarkable, unusual, and pointing us to God.

That brings me to a startling scientific realization about the Resurrection. It's the lesson of the Martian meteorites. If it *did* happen, it *can* happen.

I believe the Resurrection happened. I cannot come up with any other explanation for why the disciples insisted upon its reality, even in the face of their own martyrdom. And even more, speaking as a fan of good writing, the whole story arc of the Gospels loses its heart and goal without the Resurrection. It just doesn't work! The original version of Mark's Gospel, scholars tell us, didn't include the Resurrection appearances; they were added later, in a different style of writing. Most people assume the later addition was for theological reasons. I'd argue it was also necessary just because the story felt flat without it.

But if the Resurrection did happen, it can happen. And if it is not forbidden, then it is compulsory. My study of Mars meteorites has led me to appreciate a startling truth: the Resurrection of Jesus implies the resurrection of us all.

Sunset at Loyola Marymount University, a Jesuit university in Los Angeles.

A sunset photographed in Louisville, Kentucky.

CHAPTER SIX
WHY DO WE DO IT?

ONCE I GAVE A TALK ABOUT JUPITER AND ITS MOONS at the College of Charleston, a beautiful campus in Charleston, South Carolina. After the talk, an undergraduate came up full of enthusiasm. "I want to be a geologist!" he told me. I thought that was a fine idea. I had studied the geology of other planets, and I'd met some wonderful people in my geology department. Even if you just stick to planet Earth, you can wind up working outdoors in some of the most beautiful locations on Earth.

"Sounds great," I told him.

"Yeah," he said, "but what do I tell my mom?"

South Carolina—he was from the Bible Belt. In the culture where he grew up, studying geology and its ideas of billion-year-old rock formations directly contradicted the ways he had been taught. To be a geologist, for him, would be to declare war against his religion, his home, his family. His mom would be ashamed of him.

Scientists are people. We have moms, we have families, we have desires. Like every human being, we are a mixture of reason and heart, with hearts that have "reasons, which Reason does not know," as the French philosopher Blaise Pascal said. And like that student, we have to answer to those desires inside us and those desires inside the other people who are close to us.

There's a temptation to divide our experience into separate categories: faith versus science, emotions versus logic. But that's a false division. Real people are not just Kirk or just Spock. Heck, even Kirk and Spock were not really just "Kirk," all emotion, and "Spock," all logic. It is on the basis of reason and gut feeling that we make all the decisions of our lives. In the case of that student in South Carolina, it meant choosing between what he wanted and what he thought his family and community wanted from him, between science and his culture. People are a lot less likely to go into a field that they think will make their mother ashamed of them.

Another conversation gave me an interesting insight into the topic of science, faith, and why we do science. Oddly enough, that conversation was with Captain Kirk. How I wound up talking to William Shatner, the actor who played Captain Kirk in the original *Star Trek,* is too long a story to go into here. But when I told him I was a Jesuit astronomer, he was flabbergasted. "Wait a minute, wait a minute!" he exclaimed. As we talked, though, something became clear to me that was obvious to him but that I had never grasped before. He saw religion and science as two competing sets of truths. Two big books of facts. And what should happen if the facts in one book contradict the facts in the other?

But science is not a big book of facts.

Let me give an example of what I mean. The orbits of the planets are facts. Now, we can trace out those orbits mathematically using Ptolemy's epicycles, circles within circles encircling the Earth; or we can approximate them using Kepler's ellipses around the Sun. In fact, Ptolemy's epicycles are more precise than Kepler's ellipses: with an infinite number of epicycles, we can match perfectly the shape of any orbit. (A wag on YouTube has calculated the set of nested circles that can reproduce a drawing of Homer Simpson!) On the other hand, Kepler's ellipses are only approximately true, since they don't account for how one planet pulls on the orbit of another. But while Ptolemy's set of circles can reproduce the path of a planet, the circles themselves are meaningless except as a calculating device. Kepler's ellipses, though, led to the insight of Newton's law of gravity, which successfully predicts that orbits *should* be elliptical.

Science is not the fact that orbits are ellipses; it is the search for insights that come from those facts, understanding why the orbits should be elliptical. And it is also being open to the realization that Newton's insights are not the last word. Today Newton's law of gravity has been superseded by Einstein's general relativity, which can explain, for example, an oddity in Mercury's orbit that Newton's laws can't handle. But even relativity is not the last word. Won't the science of the year 2300 look different from what we're teaching today?

The author at the Zeiss Refractor, Castel Gandolfo.

I have never had a case in which some tenet of my faith seemed to be contradicted by what I had learned through science. (The two areas don't much overlap, to be honest.) But I do recall many times when one thing that science told me seemed to contradict something else that science told me. You rejoice in those moments, because it means that everything you thought you knew was incomplete, and you're about to learn something new.

And incidentally, our faith is not based on a rigid commitment to a book of doctrines, either. I repeated to William Shatner the phrase that the writer Anne Lamont famously used to describe faith: "The opposite of faith is not doubt; the opposite of faith is certainty." (She was actually following a famous theologian, Paul Tillich, who used a whole lot more words to make the same point.) That was completely the opposite of what Shatner thought faith was about. He'd heard the phrase "blind faith" and thought that meant accepting something as certain without looking; or worse, closing our eyes to the facts and proceeding on emotion. But that's not faith at all. Blind faith is not walking with blindfolds, blinding ourselves to truth. It's proceeding even when we can't see everything we wish we could see. We never have all the facts, and so faith is how we make essential choices anyway, about where we'll go to school, what career we'll pursue, where we'll live, whom we'll marry. All of life is making crucial decisions on the basis of inadequate or incomplete information.

Furthermore, the decisions that change our lives are often made by ignorant, stubborn, and inexperienced teenagers, like that student in Charleston. Like me. I was a goofy teenager when I decided to transfer to MIT, with some rather silly motivations: reading science fiction, exploring tunnels, weekend movies. But it determined the shape of the rest of my life.

And even after you wind up as a scientist, you still have to make decisions blindly. The hardest part of doing science is not doing experiments or calculations; it is choosing which experiments, which calculations, to perform. It's finding a topic that is big enough to be interesting but small enough that you can make reasonable progress in it with the limited resources at hand. It is deciding what science we should choose to do.

Most scientists make these decisions for practical reasons. Who will pay me? Which grants are available? Which resources do I have? What jobs are located where I want to live, where my spouse can work, where I want to raise a family or be close to elderly parents?

But for a Jesuit, the choice is centered on a different end: I am placed here to serve God and to save my soul, and everything else can be only a means to that end. That was the point of the "Principle and Foundation" found in *The Spiritual Exercises of St. Ignatius.* Indeed, in the Spiritual Exercises, Jesuits are given various ways of discerning the will of God for them.

Sometimes, discerning God's will is easy. You just know. But things are rarely that simple!

Another way of learning God's will is noticing when our soul feels consoled, even at times when we're otherwise unhappy. That, of course, was what was happening to me when I was a freshman visiting MIT. It felt right to be at MIT; it did not feel right for me when I was back at Boston College. I attributed that feeling to science fiction books or midnight tours through the tunnels under the buildings or movie nights in a lecture hall with a big screen. But I see now it was something more than that.

The third is the method of calm deliberation, as on a retreat, when we can be free from the agitations of daily life and can take the time to ponder whether one decision or another really does bring us closer to God. That's done with prayer, in which we beg God for the grace that moves our will in the direction most pleasing to him. And then we have to determine whether what we are feeling is really from God. If a stranger were in your particular place with your decision, what would you advise the stranger to do? If you were on your deathbed, what would you wish you had done?

Making decisions in this way often means that Jesuit astronomers wind up taking paths, including choices for research projects, that are different from what our lay colleagues might have chosen—paths that often our colleagues, with their real constraints of family and other commitments, cannot be free to choose.

When I got to MIT I was ecstatically happy, happier than I'd thought could ever be possible. I discovered: *this* is what I was good at. That feeling (I now recognize) was a consolation from God. I understood that MIT was where I belonged. I had finally found my passion. It wasn't the science fiction; it was the science. It wasn't the tunnels; it was the exploration. It wasn't the movies; it was the imagination. And it wasn't the science per se; it was being a part of a community of fellow nerds.

To share stories, we have to have a someone to share them with.

I would venture to say that for every Harry Potter fan who wishes they could do magic, there are a dozen who would be happy just to live at Hogwarts, to have an omniscient hat that knows where to sort you into the house where you belonged. Reading *The Lord of the Rings,* one of the longings I had was to be a part of a fellowship.

We desire community. Science is sharing stories about what we've learned of the natural world. To share stories, we have to have a someone to share them with. Eventually, if no one wants to listen to our stories about our adventures in the lab, we'll stop having those adventures. If we don't have the support of a community, it isn't going to happen. Science can't happen. Furthermore, without a community, it can't possibly be passed on to the next generation. That is why it matters to pay attention to what your colleagues think about the things you choose to do.

There are a lot of mountains and snow in India, and more people in India than all of America or Europe. But India has never won a medal in the winter Olympics. Winter sports at that level just aren't supported by Indian society. For whatever reason, very few mothers in India seem to dream of their daughters as gold-medal figure skaters. If doing science isn't the sort of thing that will make a mother proud, you're not going to find very many kids who choose to study science, much less persevere in that study. Without society's endorsement, where will we find anyone who can teach us how to do it or any students who'd hire us to teach them?

Society not only means the milieu where we grow up; it also means our society of fellow scientists once we enter the field. The work we do in our field is strongly influenced by what the other members of our field will support. That doesn't only mean what topics have grant money available. It also means whether we'll get an audience to listen to us when we present our work.

I remember the first Meteoritical Society meeting where I presented the data I was beginning to collect on meteorite densities (data that lots of people have since found good use for, by the way). One of the grand old men of the field came up to me and said, "Guy, why are you measuring meteorite densities? Nobody does that." It was a new field. It took faith—or foolishness—to do something where I couldn't be sure there would be an audience for my results.

Back when I trained for the US Peace Corps, I was one in a group of about eighty teachers, most of them right out of college and full of idealism. During the breaks between teacher training and Swahili classes we used to play a game called hacky sack. The idea is that someone tosses a bean bag into the air, and then everyone takes turns keeping it up in the air, passing it around the circle, kicking it or bopping it with your head, using anything except your hands and arms, like soccer rules. That's it. It was the kind of "cooperative," feel-good sort of activity you'd expect for a bunch of do-gooder Peace Corps types.

During stateside training, before we went overseas, the boyfriend of someone in our group came from New York City to visit. He watched us playing hacky sack, and the New Yorker couldn't figure out what the heck we were doing: "You just kick it up in the air? No goals? No contact? No trying to stop the other guy from getting it? How do you win this game?"

Well, science is like playing hacky sack. You try to keep lots of ideas up in the air, keep them alive, and pass them on to the next player. It's a cooperative activity; you gain status by how much you give, what you can add to the field, how much your work can help out the others in your field. There's never a moment when you say, "I won! You lost!"

If this is true, then what drives us to play hacky sack, to do science? And how does anyone win the game? There are plenty of obvious criteria for what a university might consider signs of success, the sorts of things that lead to tenure for someone who wants to be a professor. The approval of others, because good letters of recommendation lead to tenure. Grant money. Prizes, honors, and awards—the MacArthur Fellowship, the Nobel Prize, and other smaller prizes. Fame in general—like getting your name in the *New York Times* or *Sky and Telescope*.

Another sign of success is having successful students. It's like parenthood: sometimes we live on through our offspring. And then, of course, there's the all-time favorite metric of tenure committees: have you written papers that other people cite? The more papers, or the more citations, the more successful you are.

Those measurements might sound sort of crass, and clearly they aren't the whole picture. Another carrot out there to drive a scientist to excel is the hope of academic freedom—in other words, scientists are successful if they've reached a state where they can set the criteria for what they want to study, when they're free to choose their own topics and they're not just working for someone else. They have reached a pinnacle of success if they're not limited by what NASA will pay for but are in a position to sit on the committees that decide what NASA will pay for.

But notice what that means: the reward when we win at doing science is to be given the chance to do more science. That begs the question: what is so rewarding about getting to do science itself?

Indeed, are any of these rewards ends in themselves? For example, getting a paper in *Science*, one of the premiere scientific journals in the world, is a big deal; it's a sign of success. So do I wake up in the morning and jump out of bed to rush to the lab because, by golly, I'm going be one of the forty-odd authors on some paper that will tell the world something really new and exciting and groundbreaking about, say, the effect of collective molecular reorientations on Brownian motion of colloids in nematic liquid crystals (to cite the actual title of a paper in *Science* chosen at random)? It's important work, really; and it may be cited, oh, maybe twenty times in the next three years.

> One obvious central pillar of doing science is the search for truth.

Is that what really motivates *me,* moment to moment?

What could possibly motivate you to want to work in science? What are the itches that you have to scratch? Curiosity is one such itch. Then there's pleasure in solving problems, or merely finding patterns in the data, similar to what makes doing jigsaw puzzles and crossword puzzles so much fun. But surely science is about more than just doing a puzzle.

One obvious central pillar of doing science is the search for truth. And, of course, as we saw with the comparison of Ptolemy and Kepler, there's more to "truth" than merely getting the answer that solves the puzzle. But it might take more than a lifetime to discover if you have really found an insight into truth. Kepler didn't live to see Newton's laws. Meanwhile, is that what gets you up in the morning?

How about love? If we did not experience love in this work, we would not do it. But what is it in science that we love so much that we make it the center of our lives?

And wait a minute. How do these criteria stack up against the more practical reasons we started with? If you were on a committee looking to hire a new scientist to join your faculty, which candidate would you choose—the one who really loves the material, or the one whose papers get cited more often? And could you really support either decision?

In real life, it's never such a clean choice. Would you give up the approval of others to satisfy your own curiosity? Give up tenure to do the thing you love? Give up academic freedom, the freedom to pursue your own research, to satisfy the pleasure in solving problems? Give up fame to satisfy the pleasure in finding patterns? Give up grant money to satisfy your hunger for truth? Or vice versa?

Before you jump to one conclusion or another, remember that if you aren't getting paid then you can't afford to do the work. Without eating, you can't do any work at all. And if you aren't doing work that the culture respects and supports, then you aren't going to have very many people show up to hear or cite your paper and further that conversation we call science. You may risk facing scorn (or worse) from the people whom you rely on to support you, emotionally as well as financially.

Internal satisfaction isn't enough. Science doesn't come for free. On the other hand, without the internal motivations, without some sort of internal payout, we wouldn't do this work very long. We wouldn't enjoy it.

But I do it ultimately because I do enjoy it. That word, *enjoy,* I think, contains what I am really looking to articulate. Joy.

What do I mean by joy? Let me show you.

Thin section of the Knyahinya meteorite, photographed by the author in the Vatican Observatory meteorite laboratory.

Take a look at the photograph on the previous page showing a thin slice of a meteorite (the one here is called Knyahinya, after the Ukrainian village where it fell in 1866) seen through a petrographic microscope. The very image is joyful. What's more, and what I can only hint at here, is how joyful it is to learn the science behind the image.

Such an image is made with light that has been polarized in one orientation, which then passes through the sliver of rock (about as thick as a human hair) before it encounters another polarizing filter oriented ninety degrees from the first one. Two such filters in that orientation ought to stop all light from passing through, but the minerals in the thin slice of rock twist the polarization of the light, allowing certain colors to pass. The colors change, by the way, if you spin the thin section between the filters. The way the colors change tell us about the composition and structure of the minerals. They're also as dazzling to look at as a piece of stained glass. The round bit of green at the upper left is called a *chondrule*. To me it looks like the head of a baby in swaddling clothes, and so I sent this out as a Christmas card once.

> Ignatian spirituality emphasizes engagement with the world and "finding God in all things."

Ignatian spirituality emphasizes engagement with the world and "finding God in all things." This aligns exactly with the work of a scientist, because scientists find joy in studying things; to find joy is to find God. In addition, Ignatian spirituality provides a motivation for doing science independent of a yearning for glory or a desire for wealth. To be sure, there is not much wealth or glory to be found in astronomy!

But I have encountered fellow scientists who seem to be more interested in scoring points over their perceived rivals than in actually coming closer to the truth. More commonly, some have forgotten why they went into science in the first place and do the work just because it is the one job they are qualified for. I speak from experience. I entered the Jesuits when I was nearly forty years old and already established in my field; during the years I worked as an astronomer before I entered the Society of Jesus, I often found myself being tempted to adopt those attitudes.

Now, as a Jesuit (and with the practical advantage that I never have to worry about supporting a family), my approach toward my work has changed. I do my work out of joy, both the joy of discovery and the simple joy of becoming more intimate with this physical universe. If we believe that God created this universe, and if we believe that God so loved it that he sent his Son to become a part of it, then science becomes an act of growing closer to the Creator. In that way, it becomes an act of prayer.

A few years ago, I had a sabbatical year and taught physics at Fordham, the Jesuit university in New York City. There I had a class of really bright students taking introduction to electricity and magnetism, the very class that led to my own breakthrough when I was a student. We'd just covered Maxwell's equations. I was writing them on the board, the equations describing how electricity can give rise to magnetism and how magnetic fields can give rise to electric fields. Then, doing the mathematical manipulations that Maxwell had done back in 1865, we take a derivative here, and put in a substitution there . . .

And as I wrote down the final equation, the result of all this manipulation, a complicated scrawl of *E*'s and *t*'s and *del*'s and mu-sub-zeros, before I had a chance to turn around and explain what it all meant, my brightest student in the front row gasped, loud enough for everyone to hear him: "Oh my God. It's a wave."

Every bit of science we can extract from that slice of meteorite, or from any of the glorious astronomical images scattered throughout this book, starts with Maxwell's equations and the fact that—oh my God—it's a wave. They explain how spectral lines can tell us a star's composition and temperature, and how that information is transmitted from a star to our telescope. They explain how the polarized light can pass through a meteorite thin section in my microscope.

The fact that it's (oh my God) a wave gives us radio and television, electric power transmission of alternating current, cell phones and computers, and eventually special and general relativity. Now, it takes a couple of semesters of physics to get there, but when you do get there, take my word for it, take my student's word for it: it's an "oh my God" moment.

In my forty years of research, I've had a handful of those moments. Nothing as big as Maxwell's, of course. A couple were big enough to publish in *Science*. But it's not the final paper that I remember. It's the gasp of amazement when suddenly I saw a pattern in nature that I had not anticipated.

Awe, or "oh my God," is a very human emotion, one that could not be evoked without enough logic that we can recognize what we are seeing. It also could not be evoked if we were not so hungry for something to feed our souls that we chose to look around in the first place.

It is human to look at the stars and realize that there's more to life than what's for lunch. The very thing that makes us different from a cow or a cat, that aspect that makes us breathe "oh my God," is what we call the human soul. The soul is what the theologians tell us is the image and likeness of God. And it's why I look forward to going back to my meteorite lab. Every now and then, some unexpected result jumps out from my computer screen.

And then I go, "Oh my God."

Not all scientists think they believe in God. But I believe that they all believe in "oh my God." Science needs "oh my God." It wouldn't happen without it.

Vatican Advanced Technology Telescope (VATT) at night.

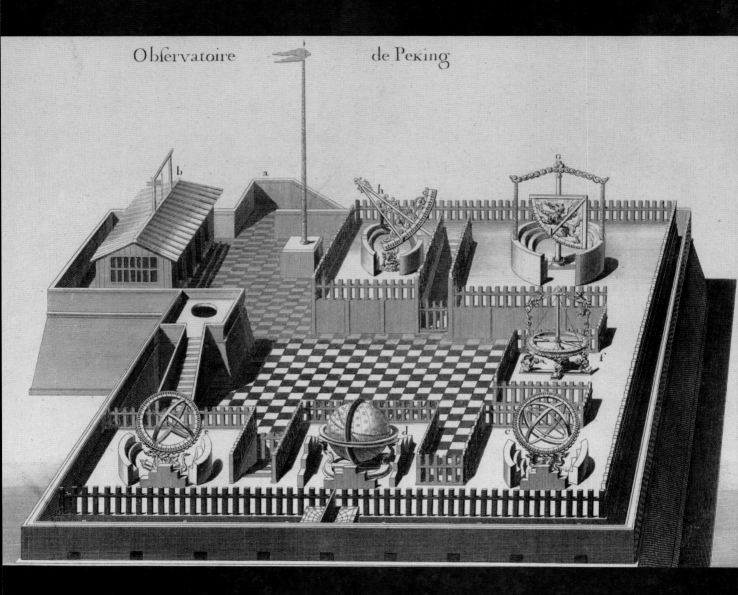

Obfervatoire de Peking

A 1674 engraving of Beijing Ancient Observatory showing six astronomical instruments similar to those that were being used in Europe.

CHAPTER SEVEN
OBSERVERS FROM THE VATICAN

I ONCE CAUSED A STIR in a little church in Hawaii. When the pastor invited any tourists present to introduce themselves and say where they were visiting from, I got up and announced that I was there because I was "an observer from the Vatican." Given that I was in Hawaii to use a telescope, I thought my description was perfectly true. I just didn't say what kind of observer I was.

Jesuits as observers (astronomical at least) are not all that unusual, but they are rarely given more than a footnote in history. That's a shame, since many Jesuit astronomers have interesting stories, too many to cover all of them here. These include the unheralded rivals to Galileo—Christof Scheiner, Johann Locher, and Orazio Grassi; the first astronomers to split double stars—Jean de Fontaney, Guy Tachard, and Jean Richaud; Esprit Pézenas, who is credited with the first observations of the Gegenschein (look it up); and Maximilian Hell, who figured prominently in the measure of the transits of Venus.

Agustín Udías, SJ, in his book *Searching the Heavens and the Earth: The History of Jesuit Observatories,* gives a summary of several hundred astronomical and geophysical observatories that were set up over the Society's 450-year history. By sheer number, they made up an impressive fraction of the astronomy and geophysics done from the sixteenth through the twentieth century. Indeed, it is important to remember that before the twentieth century, many working scientists were clergymen. After all, in those days who else had the education, the inclination, and the free time to do science?

In a later book, *Jesuit Contribution to Science: A History,* Udías gives a broader overview. The Jesuit order began almost simultaneously with the Scientific Revolution in Europe, and so it is not surprising that Jesuits, with their extensive network of schools, played a significant role in European science. But their efforts were even more important away from Europe, with scientific work done by Jesuits in mission territories. Until the nineteenth century, these Jesuits were often the only Western scientists outside of Europe, and their work in other parts of the world was essential even well into the early twentieth century. Besides their more famous contributions, such as directing the Imperial Observatory of Beijing from 1644 to 1805, they also provided important meteorological and geophysical observations in dozens of posts, from Madagascar to Manila to Havana. One sign of the extent and breadth of Jesuit science can be seen in the number of asteroids and lunar craters that bear the names of Jesuits. (A listing of the fifty-nine Jesuits so honored is found in the appendix.)

One of the earliest and most important of the Jesuits working in science was Christopher Clavius, SJ (1538–1612). He was instrumental in promoting mathematics as

A page from Clavius's 1603 book, in the Vatican Observatory collection, showing the dates of Easter and other movable feasts for the years 1984-2031.

an important part of the curriculum in Jesuit schools. He wrote math textbooks that were used throughout Europe and in mission territories. Among his achievements, it is said that he invented the decimal point and the use of parentheses to gather terms in equations. Clavius notably wrote a letter of recommendation for young Galileo, and observed Jupiter's moons through Galileo's telescope.

But he is probably best known as a part of the commission organized by Pope Gregory XIII for the 1582 reform of the calendar. The calendar reform included a new way of calculating the date of Easter and all the other movable feasts that depend on Easter. After the reform was promulgated, Clavius was commissioned to write a book explaining the reform in great detail, including tables of the dates of Easter and the other feasts up to the year 5000.

In fact, Clavius wasn't the only Jesuit to support Galileo. In May 1611, the Jesuits at the Roman College hosted a banquet for Galileo; at it he demonstrated his discoveries with his telescope. However, the rivalry for precedence in using the telescope eventually led to a falling out between Galileo and the Order.

Christoph Scheiner, SJ (1575–1650), was a professor at the University of Ingolstadt and known for his skill in mathematics. In March 1611, he adapted a telescope made by Christoph Grienberger, SJ (1561–1636), to project the image of the Sun onto a blank sheet. There he detected sunspots and measured their motions day by day as the Sun rotated. He published his results in 1612, much to Galileo's displeasure; Scheiner had not mentioned that Galileo had also seen sunspots a few months earlier. Galileo's pride of place created the first fracture in his good relations with the Jesuits. (In fact, earlier than either of them, the Englishman Thomas Harriot and the German astronomers David Fabricius and his son Johannes observed sunspots. Harriot did not publish his results, but the Germans had published theirs in a small booklet in 1611. It is doubtful that any of these early observers knew of the others' results.)

Scheiner's definitive work about the Sun appeared in 1630. In the light of our present knowledge, his descriptions of how sunspots move were far more accurate than Galileo's, noting (contrary to Galileo's assertions) that spots at different latitudes moved at different speeds. Galileo assumed the spots were on the surface of a solid Sun; Scheiner believed that the spots were parts of clouds above the Sun. Neither guessed that the Sun itself is fluid, with different latitudes spinning at different rates.

Perhaps the most notable dispute between Galileo and the Jesuits involved Orazio Grassi, SJ (1583–1654), on the nature of comets. Grassi was the first to point a telescope at the three comets of 1618; Galileo in fact never observed them. Grassi compared the position of a given comet with simultaneous observations by other Jesuit colleagues in Germany, and he determined that the comet showed no parallax; that is, its position against the background stars was the same whether seen from Germany or from Rome. He thus concluded that comets were not some local phenomenon in the sky over the observer (as Galileo had insisted) but in fact were located beyond the orbit of the Moon.

This, however, was difficult to reconcile with the Copernican system, which still assumed circular orbits around the Sun. Over the following four years, Galileo and Grassi engaged in a battle of booklets, arguing the point. Ultimately, Galileo produced *The Assayer,* his masterpiece of philosophy of science. In it he argued, with great wit and sarcasm, how one's science must rely not on the authority of some sage and only on the data. Then, on his own authority and in the total absence of data, he also asserted that comets can't possibly orbit among the planets—but that's another story.

Jesuit scientists were among those who worked the hardest to examine the Copernican cosmology from a scientific point of view. The fact was that it took more than a hundred years to address adequately the scientific problems with the idea of Earth moving around the Sun. Ultimately, this required understanding that the planets moved in ellipses, not in circles as Copernicus and Galileo thought; a new physics to explain how they moved, provided by Newton in 1687; and an understanding of the wave nature of light to explain why stars seen in a small telescope look like disks instead of points, an explanation that came only with the work of George Airy (1801–1892) in 1835. (If the stars were so close that you could see them as disks, it should have been easy to watch them shift their positions as Earth moved around the Sun. In fact, their great distance made such measurements beyond the ability of seventeenth-century telescopes to detect.)

Johann Georg Locher, SJ (1592–1633), a student of Scheiner, published *Mathematical Disquisitions on Controversial and Novel Astronomical Topics* in 1614. It attacked the Copernican system on scientific grounds. Galileo mocked the book in his own writings; while Locher was ultimately incorrect, Galileo's sarcastic arguments against him were equally faulty.

Giovanni Battista Riccioli, SJ (1598–1671), wrote *Almagestum Novum* in Italy fewer than twenty years after the Galileo trial. In it he outlined more than eighty arguments for and against the Copernican system. Among other arguments, he realized that if Earth were spinning, a cannonball shot to the north would be subject to what we now call the Coriolis force, named after Gaspard-Gustave de Coriolis (1792–1843), the nineteenth-century scientist who finally described it mathematically. Since the actual speed of a spinning spherical Earth changes slightly as latitude changes (because the equator has the farthest to rotate, while the poles don't move at all) this should make a cannonball shot to the north appear to bend eastward as Earth rotated underneath it. Since no one had ever seen such a deflection—which in truth is far smaller than all the other things that could affect a cannonball's flight—he concluded that Earth did not spin! Brilliant science, wrong conclusion.

However, what is most remembered is Riccioli's famous map of the Moon, shown on the following page. His was not the first map of the Moon, but it was certainly the most detailed and accurate of its era. In it he devised the nomenclature we still use today. These include some two dozen craters named for Jesuits, including himself and his student Grimaldi. But he also named prominent craters on the Moon after Copernicus, Kepler, Galileo, and Aristarchus, who all supported the heliocentric system. More importantly, the system of nomenclature he devised set the pattern for how we name features on the surfaces of all the other moons and planets, especially now that spacecraft have revealed so many different planetary surfaces.

Riccioli ultimately disagreed with the Copernican system: his arguments were founded not on some simplistic reading of Scripture but on specific scientific challenges. In some cases, it would take centuries of scientific advances before the questions he raised could be answered adequately. But, of course, that's the power of science: its willingness to challenge even attractive ideas, and its ability to learn and advance from seeing how those challenges turn out. That, more than any detail about how planets moved, was the real scientific revolution.

While the new astronomy was being found controversial in Europe, it was being used by Jesuit missionaries in China as a way of opening that nation to Western ideas.

When it comes to missionaries in the scientific era, few are held in such renown as Matteo Ricci, SJ (1552–1610). He studied mathematics and astronomy with Clavius

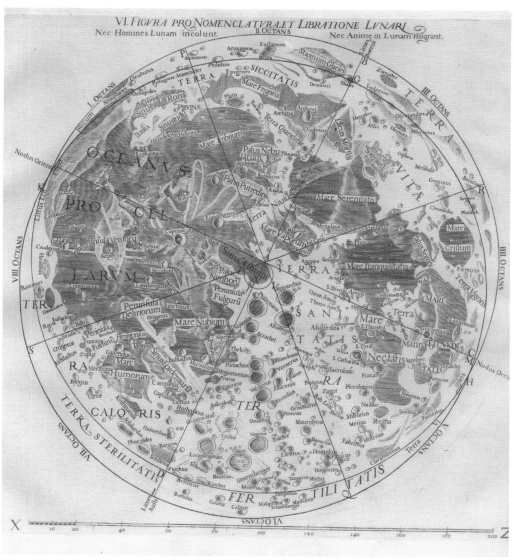

Riccioli's map of the Moon in his *Almagestum novum*, from the Vatican Observatory's collection.

at the Roman College, as well as philosophy and theology. In 1578 he left Europe for India, and then in 1582 he continued to Macao, where he learned Chinese. A year later, entering China itself, he discovered that his mathematics and astronomy marked him as a scholar of respect among the Chinese, given their own long history of astronomical observations. In January 1599 he finally reached Beijing.

While Ricci made his name among the Chinese as a mathematician, other Jesuits traveling with him introduced the new Western astronomy to the Chinese court. In 1610, the Chinese astronomers had made a serious error in predicting a solar eclipse, while Sabatino de Ursis, SJ (1575–1620), an Italian Jesuit astronomer who had arrived in China to help Ricci, predicted the eclipse correctly. After the failure of Chinese astronomers to predict another eclipse in 1629, again correctly described by the Jesuits, the missionaries were given the command of reforming the Chinese calendar.

Other Jesuits introduced the telescope and Galileo's recent discoveries. Among those doing this work in China were Johann Adam Schall von Bell, SJ (1592–1666), who also had studied at the Roman College while Clavius was still teaching there and at the time Galileo demonstrated his telescope. Arriving in China in 1619, during a period of political turbulence, he worked on the calendar and other astronomical observations. Meanwhile, the accuracies of the Jesuits' eclipse predictions continued to impress the Chinese. Finally, in 1645, Schall was named the first Jesuit director of the Imperial Observatory, with a rank of mandarin. As Udías comments in his book: "Schall took part in important state decisions, even in the choice of the crown prince. In the whole history of China, probably no Westerner ever had the influence that Schall had during the life of Emperor Sunzhi."

Ferdinand Verbiest, SJ (1623–1688), arrived in Beijing in 1660 and served as Schall's second in command and closest collaborator. When the displaced Chinese astronomers, under a new emperor, conspired to have Shall, Verbiest, and other Jesuits arrested in 1665, the Westerners once again used eclipse predictions to support the superiority of their new astronomy. Nonetheless, Schall was sentenced to death. But a major earthquake occurred the day following his sentencing; this convinced the Chinese that the sentence was unjust, and Schall was released. Unfortunately, he died of ill health the following year. Verbiest took over as director of the Imperial Observatory, a post that a Jesuit would hold until 1805.

China was not the only place to receive Jesuit missionaries with an interest in astronomy, of course. In 1686 a group of seven French Jesuits scientists and a large amount of scientific equipment left France for elsewhere in Asia. Jean de Fontaney, SJ (1653–1715), was the head of the expedition; stopping en route in South Africa in 1685, he and Guy Tachard, SJ (1651–1712), observed the brightest star in the Southern Cross, Acrux, from the Cape of Good Hope and discovered that it was in fact a close pair of stars—a double star. They continued their astronomical work after they reached Siam (Thailand). In a similar way, Jean Richaud, SJ (1633–1693), observing in 1689 from Pondicherry, India, was the first to split Alpha Centauri.

Roger Boscovich, SJ (1711–1787), was probably the most outstanding Jesuit scientist and engineer of the 1700s. The sixteenth Jesuit General Congregation, held in 1731 (when Boscovich was twenty years old), had emphasized the need for Jesuit schools to support the Aristotelian system as desired by authorities of the Catholic Church at that time, since Aristotle's philosophy was the basis of the prevailing Catholic system of theology. But the General Congregation also opened the door to the possibility of teaching

Late seventeenth-century watercolor titled
Observation of an Eclipse of the Sun from a Jesuit Observatory in Siam.

experimental sciences. Soon Jesuit schools used that as license to teach Newtonian physics, and so Boscovich was trained in that system. Showing his interest in the modern sciences, he observed in 1736 the transit of Mercury across the face of the Sun.

As a young man, Boscovich achieved his first fame in Rome in 1742 by proposing and directing the repair of cracks in the dome of St. Peter's. He was among those who worked hard to have the Vatican lift its prohibition against teaching heliocentrism, which finally occurred in 1757. The ceiling of the mathematics room at the Jesuit College in Prague, dating from 1760, shows little putti looking through telescopes at stars, each star surrounded by planets and comets.

Of his many works, the most important was *Theoria philosophiae naturalis,* published in 1758, and generally credited with the origin of the modern atomic theory of matter. But also notable among Boscovich's activities was promoting observations of the transits of Venus in 1761 and 1769. This was perhaps the most important scientific achievement of the eighteenth century. By measuring the path of Venus across the Sun as observed from many different, precisely measured locations on Earth, one could triangulate the distance between Earth and Venus. From that, one can calculate the distance from Earth to the Sun and the other planets. Eventually, on the basis of that measurement, one could calibrate the other rungs of the "distance ladder" to the stars and galaxies on which the Hubble scale and the size and evolution of our universe can be understood.

At the time of the transits, 25 percent of all observatories in Europe were run by Jesuits. One of the most prominent of these was the Vienna University Observatory, founded by Maximilian Hell, SJ, and built between 1753 and 1754. In 1769, Hell led a Venus transit expedition to Lapland, at the invitation of the king of Sweden, at a time when it was otherwise illegal for Jesuits to set foot in Sweden!

But in 1773, when the Jesuits were suppressed, his observatory was absorbed by the state. In the following decades its management was turned over to lay people, who were often very anticlerical. Thus, some fifty years later, in 1835, the then director of the observatory, Joseph Johann von Littrow (1781–1840), looked over the notes that Hell had left behind and accused him of falsifying his data!

The publication of Hell's observations of the transit had been delayed for all the usual reasons, and thus they came out after the first results were already calculated from other observations. Littrow concluded that Hell's observations were too good to be true; indeed, Littrow announced that the ink used in some entries was of a different color, which indicated to him that some numbers were written into his observation notebooks at a later time. This became the received wisdom of the early nineteenth century: Hell was a fake. But then, he was also a duplicitous Jesuit; what could you expect?

That was the fate of his reputation until 1883. That year, the eminent Canadian-American astronomer Simon Newcomb (1835–1909) wrote up the case in the *Monthly Notices of the Royal Astronomical Society*. Examining the pages in question that Littrow had cited as evidence of Hell's malfeasance, Newcomb wrote that he was "perplexed to find himself differing entirely from the conclusions of Littrow." Not only did the numbers seem authentic to him, Newcomb eventually discovered that Littrow himself was, in fact, color blind.

There is a certain irony here since Newcomb himself was often accused of trying to destroy the careers and reputations of his rival scientists. Possibly he enjoyed pulling Littrow down a peg, even if it meant defending a Jesuit. It is said that another of Newcomb's targets was a friend of Arthur Conan Doyle, who then used Newcomb as the model for Sherlock Holmes's nemesis, the evil Professor Moriarty.

By the end of the nineteenth century, Jesuit science had taken on a new apologetic role within the Catholic Church. Up to then, scientific work had been seen as a normal part of the activities of the clergy. But by the 1870s a number of pressures within both secular society and the church had put a strain on the relationship between science and religion.

The idea of a conflict between science and religion was an outcome of Enlightenment anticlericalism and its expectation that science could in some way replace religion, and the church's sometimes uncreative and stubborn response to Enlightenment ideas. The rapid development of technology in the years that followed contributed to the "Whig" myth of the inevitability of progress and its expected triumph over older ways of thought.

Both trends were reflected in the rise of the nineteenth-century secular university, especially in Germany, and the evolving status of science as a professional occupation. In Italy the anticlerical government, which had conquered Rome in 1870 and "imprisoned" the pope in the Vatican, built a monument to the sixteenth-century self-styled magician and charlatan monk Giordano Bruno, turning him into a martyr of modern science.

This development was tied to the burgeoning eugenics movement, which attempted to use Charles Darwin's ideas of evolution as the basis of breeding "superior" human beings and eliminating the "unfit" by forced sterilization. When the church spoke out against eugenics, it was labeled as "against progress." In America, Andrew Dixon White's *A History of the Warfare of Science with Theology in Christendom* contributed to the movement for blocking immigration from the Catholic nations of southern and eastern Europe because Catholics were obviously antiscience.

One of the most prominent Jesuit astronomers of the era was Angelo Secchi, SJ (1818–1878), the director of the Roman College Observatory. We'll learn more about him in a later chapter; what suffices here is that the popes noticed Secchi's international repute in science, and so after his death and after the Italian government had confiscated his observatory at the Roman College, Pope Leo XIII looked to one of Secchi's assistants to set up a new Vatican Observatory within the walls of the Vatican.

The first work of the newly reformed Vatican Observatory was to join in an international astronomical project, the eighteen-nation Carte du Ciel photographic atlas of the stars, which began observations in 1891. Among other benefits, participation in this project gave the Holy See de facto status as a nation independent of Italy.

The first members of the Observatory were not Jesuits, but in 1906, Johann Georg Hagen, SJ (1847–1930), was named director of the Observatory. Hagen, an expert in double star observations, was an Austrian who had immigrated to America and had previously directed the Georgetown University Observatory. After the reconciliation between Italy and the Vatican in 1929 and Hagen's death in 1930, the Observatory moved out of Rome to the papal summer palace in Castel Gandolfo, and the staffing of the Vatican Observatory was entrusted entirely to the Jesuits. At that time, a group of young Jesuit physicists set up a laboratory to measure the spectra of pure metals, which could be used to compare against the observed spectra of stars. In the 1940s, they founded the journal *Spectrochimica Acta,* which was printed at the Vatican in the years immediately following World War II.

The Vatican Observatory today has a dozen Jesuit researchers with doctorates from around the world who work in a wide range of fields. It also continues to play an active role in international programs and organizations, such as the International Astronomical Union. For example, the chair of the resolutions committee that wrote the modern definition of Pluto as a dwarf planet was Rev. Christopher J. Corbally, SJ.

So what is distinctive about Jesuit astronomy? Doing astronomy as a Jesuit comes with remarkable advantages. But every advantage has its matching cost.

Young Jesuits, regardless of how rich or poor their families when they enter the order, are all given the opportunity and support to pursue advanced education. But a typical Jesuit is also committed to extensive studies in philosophy and theology. Thus, a science PhD is often either interrupted or delayed.

Notice how Grassi's network of collaborators allowed him to compare the positions of comets observed from Rome with those by fellow Jesuits in Germany, something Galileo could not do. And Jesuit missionary scientists were sent to exotic places around the

Johann Georg Hagen, SJ, director of the Vatican Observatory from 1906 to 1930, at the 1909 Historic Specola Hagen Telescope (c. 1925).

world, with their transportation and support paid for by the church. But being a missionary also meant that they were physically removed from communities of scholars or up-to-date resources, and missionaries also had nonscientific duties that would call them away from doing science.

While being a Jesuit offered instant credibility in some circles, it led to instant dismissal in others. For example, Secchi was exiled from Rome for two years during the revolutions of 1848 simply because he was a Jesuit, and he fought significant prejudice from anticlerical Italians and anti-Catholic British scientists.

Finally, the expectations of the church can put significant strains on how Jesuits present their work. Jesuits are seen as speaking for the church; thus, it is important that when Jesuits present their results, they do so in such a way as to avoid bringing scandal to the church.

My own scientific work follows this pattern. I work with meteorites. Brother Robert "Bob" J. Macke, SJ, and I measure the density and porosity and other physical properties of every type of meteorite, and once we have about a thousand data points, we can begin to look for patterns. We can do such a job only because we are Jesuits and working for the Vatican Observatory. Beyond the unusual access to the Vatican's large collection of meteorites, it took ten years of developing and practicing our technique before we had enough data to show the world why the numbers mattered. No young scientist on a three-year grant cycle or a six-year tenure deadline would have had the time needed to do this work. We were also able to take our equipment to meteorite collections around the world; the Vatican name opened doors, and the presence of local Jesuit communities where we could stay made it affordable. The importance of this work is seen now in the fact that Brother Bob is in high demand for his data and his expertise. He is a participating scientist on the recent NASA mission OSIRIS-REx, for which he is measuring the samples brought back from asteroid Bennu, and he is a part of NASA's Lucy mission to explore asteroids co-orbiting with Jupiter.

But we also face some of the same disadvantages as earlier Jesuits. When Brother Bob joined our meteorite group in Rome, he had a doctorate in physics and three master's degrees—which meant that he was forty years old before he was finally able to set up his own lab, ten years later than typical non-Jesuit contemporaries.

Finally, while none of our work has led to the possibility of scandal within the church, we are very aware that if we cut corners or act inappropriately or just publish bad data, it will reflect badly not only on us but also on the Jesuits and our church. We have to hold our behavior to the highest standard. And in spite of our best efforts, people do love to try to bring scandal to us! A search on the internet turns up all sorts of sites claiming wild things about us. Supposedly, we have a telescope named for the devil, and I am purported to have said some pretty crazy things about aliens and UFOs.

In addition, the sort of science that Jesuits have been free to do, such as cataloging and amassing data over a long period of time, often means that our work is unappreciated by some of our immediate peers. Famously, in the nineteenth century the Whig historian and politician Thomas Babington Macaulay sneered that being a Jesuit "has a tendency to suffocate, rather than to develop, original genius."

This goes to the final issue of what it means to do science as a Jesuit: the motivation behind the work. The unspoken assumption of someone like Macaulay was that one does science for the glory, or the recognition of "genius," it brings upon the scientist. But to a Jesuit, the glory that comes from the science ought to be the greater glory of God the Creator, not glory for the person who happens to have revealed some detail of that creation, much less the one who managed to get the grant.

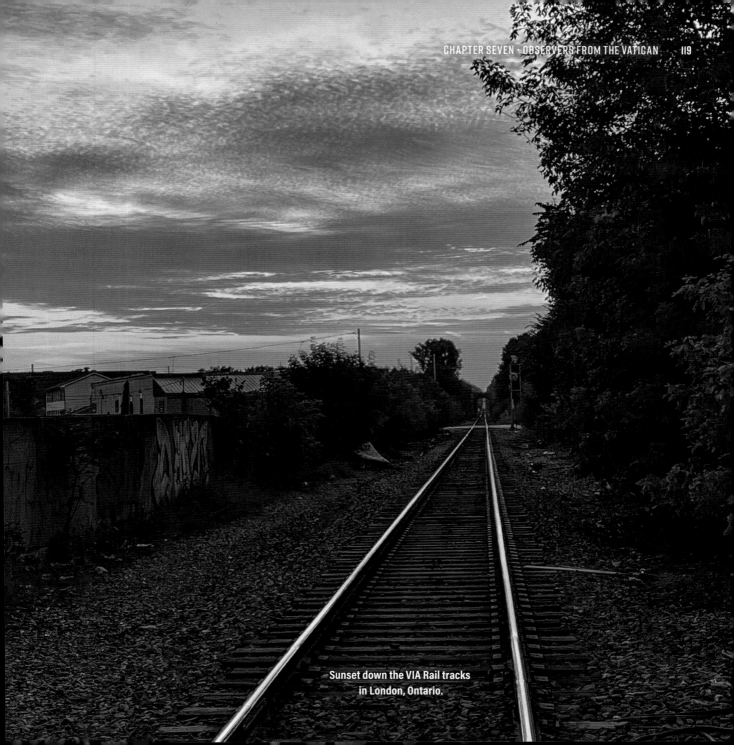

Sunset down the VIA Rail tracks
in London, Ontario.

A portrait of Angelo Secchi, SJ.

CHAPTER EIGHT
THE FATHER OF ASTROPHYSICS

ANGELO SECCHI, SJ, A NINETEENTH-CENTURY JESUIT PRIEST from a small town in northern Italy, might be the greatest scientist most people have never heard of.

True, Secchi is not completely forgotten. If nothing else, he is known today for two instruments that bear his name: the Secchi disk, which he invented as a favor for a friend looking for a repeatable way to measure the clarity of ocean water; and the Sun Earth Connection Coronal and Heliosphere Investigation (SECCHI) package, an instrument on a NASA spacecraft monitoring solar activity. What's better fame than to become a NASA acronym?

But he accomplished so much more than that. As a physicist working with the Holy See, which was responsible for the civic government in the area around Rome in the mid-nineteenth century, he provided practical applications of science to many aspects of ordinary life, such as timekeeping, the proper drainage in cities, and establishing standards for reinforcing buildings against earthquakes. He built the first electromechanical device to automatically record meteorological data, and he set up a system of weather reporting using the telegraph. He helped to found the first magnetic observatory in Italy, investigated the origin of hail and other meteorological phenomena, installed sundials and lighthouses in the Papal States, and surveyed the Appian Way. He also played a key role in the revision of the metric system. But mostly, he was an astronomer. And most crucially, more than anyone else of his era he invented the fields we now call astrophysics and planetary sciences.

It is a shame that Secchi is not better known. His life and personality were as rich and colorful as Galileo's, maybe even more so. For example, Galileo, after his first

Sunset after a storm in Englewood, Florida.
Secchi pioneered the modern study of meteorology.

discoveries with the telescope, spent the rest of his career writing works of philosophy and popularization. Secchi was also well published in popularization and philosophy, but he continued to produce profound advances in science in half a dozen widely divergent fields.

Galileo, for all his famous problems with church authorities, was never forced into exile from his homeland. Secchi was driven out of Italy once, and nearly had to flee a second time, given the agitations of his enemies in the highest levels of the anticlerical rulers of Italy.

Galileo was supported by two of the wealthiest and most powerful city-states of his era, first Venice and then Florence, and he operated at the center of the scientific world of his era. In contrast, Secchi had to work with the budget of the Roman College, limited compared to his rivals, who had all the resources of the national observatories of France and England. And Secchi worked and often wrote up his results in Italian, a language that many of his contemporaries did not read. Indeed, his rivals in the United Kingdom, especially Norman Lockyer, the editor of the important scientific journal *Nature,* made sure that Secchi's most important works were never translated into English.

Before Secchi, astronomy was only "astrometry," the study of the positions of the stars and planets. This meant merely carrying on the same program of Tycho Brahe from back in the sixteenth century, which itself was pioneered by the ancient Greek, Hipparchus. Astrometry was very practical: the ancients wanted to be able to predict planetary positions to improve their astrology, while the national observatories of Secchi's time wanted better star positions for timekeeping and commercial navigation. But Secchi was the first person to think of the stars in the sky as places with their own compositions and histories. He was curious about them not for commercial reasons, but for their own sake.

The comparison of Secchi and Galileo can also extend to their combative person-alities, with which they reacted to the turbulent currents of the politics of their times. Secchi was subjected to no shortage of slander.

One final comparison: Galileo died quietly in his home in Arcetri, at the age of seventy-seven. Secchi died at the much younger age of fifty-nine, his health worn down by the stresses of his travels and his political fights. Of course, if Galileo had died at fifty-nine, he'd never have had to endure his infamous trial.

In the summer of 2023, I had a chance to see a Secchi disk in action. Dr. Geoffrey Schladow, the director of the Tahoe Environmental Research Center (TERC) at the University of California, Davis, invited me to come and give a public talk about Angelo Secchi at its Lake Tahoe center, and as part of the deal, I got to go out on the lake and see how the disk works.

The Secchi disk is a ridiculously simple but remarkably powerful way of getting accurate, repeatable measurements of the clarity of waters in the ocean, large lakes, rivers, streams—wherever you want to measure the amount of dirt or algae or other pollutants you might find in natural waters, and more importantly, how they may have changed with time.

The idea is indeed simple. You attach a white disk (sometimes with black stripes) to a rope, lower it in the water, and then measure how deep it has to go before the disk can no longer be seen. The more rope you need, the clearer the water. There's not much training required, just a few tricks to make sure you aren't blinded by sunlight glinting off the surface of the water. And as for equipment, it's nothing more complicated than a measured rope and a simple round disk.

The author with a Secchi disk on Lake Tahoe.

Anyone can use a Secchi disk. And almost everyone has! It's a perfect opportunity for ordinary folks to get involved in useful scientific measurements, or what's come to be called citizen science. See, for instance, Secchi Dip-In, a volunteer program started in 1994 (secchidipin.org). Today you can find online millions of Secchi disk-depth readings from around the world. In some locations, the readings cover a time span of more than 150 years. Remarkably, the results don't much depend on who is making the measurements, or whether the sky is clear or cloudy, or whether the sea is calm or stormy.

So how did a Jesuit astronomer get his name attached to those measurements?

Alessandro Cialdi (1807–1882), who trained as an engineer, served as commander general of the Pontifical Navy back in the 1860s when the Holy See was still a recognized nation that encompassed Rome and its environs—more about that later. Cialdi was particularly interested in the transparency and color of water and how that might change relative to the height of the Sun in the sky, the weather, sea conditions, and so forth; from this he looked to derive the correlation between transparency, waves, and currents, especially in the seas around Civitavecchia, the principal port serving the city of Rome. In 1865, Cialdi invited Secchi to join him for a measurement campaign aboard the Pontifical Navy's steam corvette *Immacolata Concezione*. From April 20 to June 1, they measured the state of the waters using some of the most advanced instruments of the day.

The idea of lowering something into the water to see how deep it could go before it was obscured wasn't new. In the past, captains had been known to put a dinner plate into a net and lower it down. But Secchi approached the problem systematically. He tried disks of different colors and sizes: the first was over twelve feet in diameter, an iron ring covered with a canvas painted white, then smaller disks (around fifteen inches) of various colors. He worked out a system of weights to keep the disks horizontal in the water and used markings on the ropes to determine the depth. And then—most critically—Secchi and Cialdi wrote up their results and published them in the journal of the French Académie des Sciences.

Around 1900, George C. Whipple (1866–1924) added the black-and-white bits often seen on these disks, although all-white disks continue to be used in the open sea (and on Lake Tahoe). About the same time, François-Alphonse Forel (1841–1912), the founder of the science that studies lakes, called limnology, named the disk in honor of Secchi.

So who was Secchi, and how did he get to be invited on this cruise in the first place?

Angelo Secchi was born in 1818 into a middle-class family in Reggio Emilia, northwest of Bologna. (If you are not familiar with the Don Camillo stories of Giovannino Guareschi set in Reggio Emilia, dig them up—they're a wonderful window into the world of postwar Italy, wry and funny and wise.) Secchi entered the Society of Jesus in Rome at the age of fifteen. He did well in school, and by the time he was twenty-three, he was teaching physics at the Jesuit's Collegio Illirico in Loreto. By 1844 he had gone back to the Roman College for theological studies and was ordained a priest in 1847, at the age of twenty-nine.

All these events occurred during turbulent political times in Italy. In particular, 1848 was a year of tremendous upheaval throughout Europe. In November, as Garibaldi's armies entered Rome, Pope Pius IX was forced to flee to Naples, and the Jesuits were expelled from Rome by the new Roman Republic.

In exile, Secchi spent six months at Stonyhurst College in the north of England, where he may have worked at the

View of Fr. Secchi's observatory on the roof of St. Ignatius Church, as seen from the roof of the nearby Roman College (c. 1865).

astronomical observatory. By 1849 Secchi had moved to Georgetown University, in Washington, DC, where he taught astronomy and learned the new and then controversial ideas about meteorology from Matthew Fontaine Maury (1806–1873), the superintendent of the US Naval Observatory and head of what was then called the Depot of Charts and Instruments.

In July 1849, supported by the French army, Pope Pius IX was reinstalled in Rome. The following year, Secchi returned to the Roman College, where he was named director of the observatory there. He was thirty-one years old.

His first achievement was to fulfill a dream of the famous Jesuit scientist Roger Boscovich and install a modern observatory on the flat roof of the church of St. Ignatius in Rome, atop pillars designed to carry the weight of a dome that had never been built. It was there that Secchi completely changed the nature of the questions that astronomy would ask.

Secchi began by studying the Sun. He went on an expedition to photograph the solar eclipse in Spain in 1860, the first eclipse to be photographed. When he compared the corona in his photos with that seen in photographs taken by Warren la Rue (1815–1889) at a different location, he deduced that the fact that the corona was the same in both photos showed that it was not the effect of Earth's atmosphere but a real part of the Sun. Back in Rome, he went on to make remarkably detailed observations of sunspots and solar spicules. (See photograph on page 130.)

At the same time, he made regular records of the direction and strength of Earth's magnetic field. This work eventually led to the recognition of the connection between solar activity and changes in this field, hence his name on the SECCHI package.

But his most remarkable achievement was observing the spectra of stars and classifying them by their spectra.

A little bit of background is in order. In the 1820s, Joseph Ritter von Fraunhofer (1787–1826) had taken spectra of the Sun and the bright star Sirius, and he had noted that they were distinctly different. But this work was neither understood nor followed up on at the time.

MINUTES OF ARC.

Mem. Roy. Astron. Soc. Vol. XLI, Plate 2.

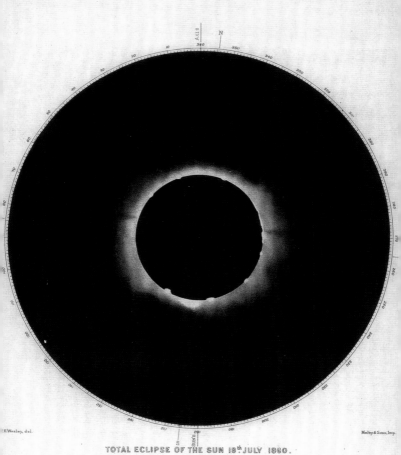

H. Wesley, del.

Malby & Sons, Imp.

TOTAL ECLIPSE OF THE SUN 18th JULY 1860.

The above drawing has been made by combining the details visible in photographic copies (on paper)
of four negatives obtained at Desierto de las Palmas by M. Montserrat, with the telescope of Padre A. Secchi.

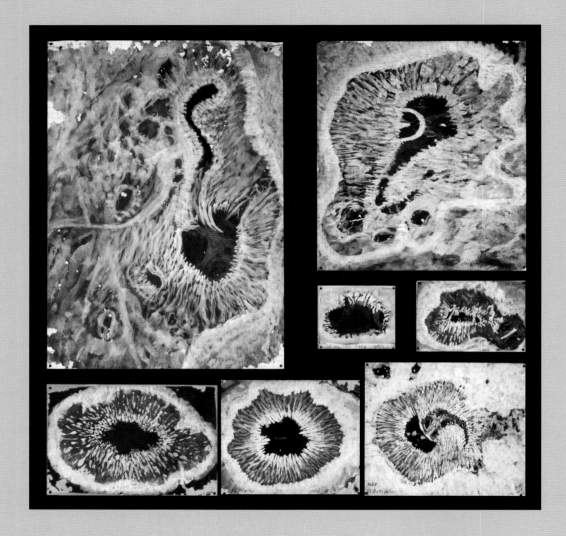

Spicules (thin lines surrounding sunspots) detailed by Fr. Secchi from his
observatory at the Roman College.

Instead, the common wisdom was that astronomy was the study only of the positions and motions of the stars. For example, in 1832, Friedrich Wilhelm Bessel (1784–1846), the mathematician who gave us Bessel functions, had written: "What astronomy must do has always been clear—it must lay down the rules for determining the motions of the heavenly bodies as they appear to us from the Earth. Everything else that can be learned about the heavenly bodies . . . is not properly of astronomical interest." And indeed, for that reason astronomy was often considered a branch of mathematics.

In 1835 the French philosopher Auguste Comte (1798–1857) insisted that astronomy provided a perfect example of the kind of knowledge that humans know must exist and yet could never know: "Every research in relation to stars not reducible in the end to simple visual observations is perforce barred to us . . . we could never study, by any means, either their chemical composition or their mineral structure. Our positive knowledge concerning the stars is necessarily restricted just to their geometrical and mechanical phenomena, being it entirely impossible to undertake any physical, chemical or physiological research."

Going to a star and fetching a sample to measure in our labs would be impossible, right? So how could we ever know what they were made of?

In 1859, Gustav Robert Kirchhoff (1824–1887) and Robert Wilhelm Eberhard Bunsen (1811–1899) showed how the missing dark lines in the spectra could be related to the chemical composition of the gases that absorbed that light.

Secchi followed up on this work by putting a large prism in front of his objective lens so that every star he observed became a spectral rainbow. And then he systematically observed five thousand stars, classifying each and every one by its spectral lines.

TAV. III.- Tipi diversi degli spettri stellari

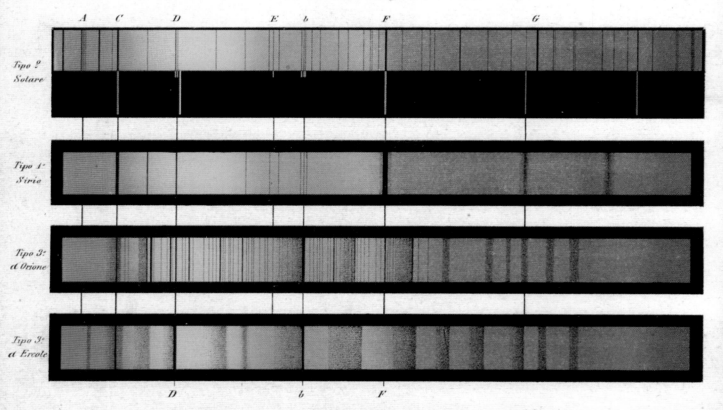

Micheletti - ROMA

He noted the differences between stars whose spectra were filled with lines and those lacking lines; and he noted those whose line positions were significantly different from the Sun's but identical to the lines of carbon seen in the lab—what we now call carbon stars. (Some of Secchi's spectral rainbow drawings are shown on page 132.)

Why was this significant? What could this tell us? Secchi realized what it would mean; in some ways his understanding was even more impressive than his work on those groundbreaking observations. In 1863 he wrote:

> Spectral studies of celestial bodies are not aimed just at curiosity, but on them depends the solution of many important cosmic questions.
>
> The first and the most important one is to recognize the nature of the matter composing the atmosphere of the celestial bodies . . . another problem whose solution can be sought through spectral analysis is that of the proper motions of many stars. . . . A third goal could be reached through this research, that of fixing the proper colors of the stars and their corresponding brightness. . . . Lastly, spectral analysis can be used also for photometric studies.

All of these results, and more, have since come out of this work of spectral classification. The classification of stars eventually would lead to the Hertzsprung-Russell diagram, which forms the basis of our understanding of stellar evolution. Secchi changed the question of astronomy from Where are the stars located? to What are the stars made of, and how did they get to be that way? And the use of spectra to measure the motions of stars is what made possible the first discovery of planets orbiting other stars, and it's the foundation of the work by Hubble and Lemaître that we now call the Big Bang!

But notice another thing about this work. Remember how, with the Secchi disk, other people had thought to lower a disk into the ocean, but it took Secchi to turn this trick into an actual science? The same pattern is at work here. Others had noted that the Sun had a particular spectrum; Secchi used spectroscopy to answer specific questions about the solar corona. Others had noted that spectra were visible in bright stars; but Secchi decided to measure the spectra of thousands of stars, looking for patterns and ways to classify them. Furthermore, he also took the spectra of planets, identifying chemicals in their atmospheres.

Secchi continued this pattern of asking where and also what with his observations of the planets. Larger telescopes than Secchi's had existed since the end of the eighteenth century, and Mars makes an especially close approach to Earth every twenty-two years or so. But few people before Secchi had bothered to actually try to inspect the surface of Mars and make sense of its surface markings in terms of geological processes. (See page 135 for examples of Secchi's pioneering illustrations of the surface of Mars.) Secchi's work inspired the pioneer maps of Giovanni Virginio Schiaparelli (1835–1910) at the next favorable conjunction, twenty-two years later. Schiaparelli and Percival Lowell (1855–1916) infamously borrowed Secchi's term *canale* to describe features that turned out to be optical illusions. In contrast, the features that Secchi described as "channels"—*canale*—are real.

At the 1867 Paris Exposition, which included exhibits from every nation in Europe, including the Holy See (not yet occupied by Italy), Secchi demonstrated an automatic weather-recording machine, the meteorograph, which won the fair's gold medal. At that time Secchi was awarded the Légion d'honneur by Napoleon III. That same year, the French formed a commission with national representatives to vote by nation on various proposals to redefine the metric system, and Secchi was the obvious person to represent

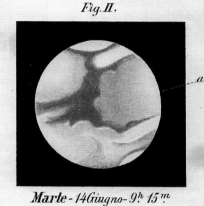

Fig. I.

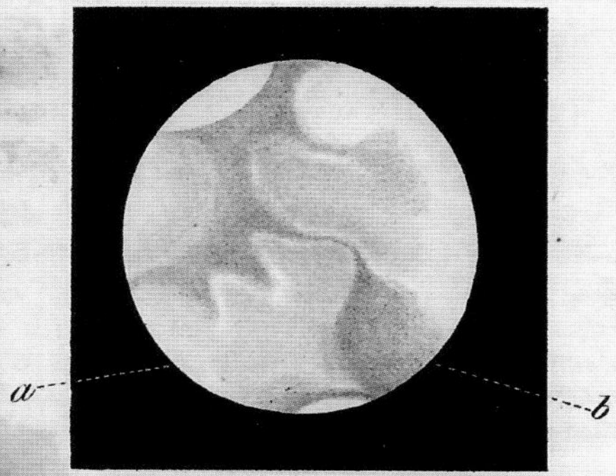

Marte - 3 Giugno - 9.h 45.m p.m.

Fig. II.

Marte - 14 Giugno - 9.h 15.m

Zu den Astronomischen Nachrichten N.o 1157.

Chromolithogr. v. W. Nathansen, Graveur, Hamburg.

Secchi's illustrations of the surface of Mars, published in 1858 in the journal *Astronomische Nachrichten*, and scanned at the Vatican Observatory Library

the Holy See. In 1870, out of that commission the Committee of 1870 was formed, with Secchi nominated to play a key role.

Then Germany invaded France. The French withdrew their army from Rome to defend Paris, and that allowed Garibaldi to march once again into Rome. Newly unified Italy declared that the Holy See no longer existed as a nation. But the pope, insisting that the walled area around St. Peter's was still independent of the Italian government, declared himself "a prisoner in the Vatican"—an issue that would not be resolved until 1929.

In 1872, the Committee of 1870 met again in Paris and expressed the wish that "in spite of the changed conditions of the Holy See, the International Commission would not be deprived of the personal contribution and studies of Fr. Secchi, which were of great benefit to the work of the Committee in the previous session."

This was a direct rebuke to the Italian government, which had tried to have Secchi removed from the work because he represented the Vatican as a nation that, in their eyes, no longer existed. If Secchi were allowed to continue his work, it would be a de facto recognition of the Holy See as independent of Italy. The Italian minister plenipotentiary in Paris, Costantino Nigra, sent a telegram to the French foreign minister: "Fr. Secchi was sent here to take part in the work of the Commission of the Meter. He was received as a representative of the Holy See. Please, make known to me if there is a place to ask for explanations from the French government . . . to express our reservations."

The Italian scientists tried to explain to their government that this action "would produce a bad impression on the Commission and the learned world." But the Italian foreign minister insisted that "the Government of the King cannot let pass in silence the designation, made in the French Official Gazette, of a State that no longer exists in

European public law." He ordered the Italian delegates to protest and to refuse to take part in any decision of the commission.

The rest of the commission disagreed, and members voted to let Secchi continue as a representative of the Holy See. In the voting, the Italians abstained. Following the vote, Italy was excluded from the Standing Committee.

This was noticed back in Rome, where it was considered a personal victory of Secchi and an indirect political achievement for the Vatican. Once he returned to Rome, Secchi was received in audience by Pope Pius IX. In his diary, he recorded: "The Pope welcomed me with a hand raised, saying 'I vote for Fr. Secchi.'"

When Garibaldi's armies marched into Rome, Secchi's university, the Roman College, had been confiscated by the new anticlerical Italian government. Secchi gave up his chair of astrophysics at the Roman College and refused the positions and honors offered to him by the new government—they were hoping to set him against the pope's claims of independence. With the expropriation and confiscation of church properties under the new regime, Secchi assumed he would lose the use of the Roman College Observatory. But his reputation was such that he was allowed to continue his work, and in 1876 the government confirmed him as director of the Observatory.

Alas, just two years later, in 1878, he died of stomach cancer. A note in Schiaparelli's observing notebook on the night of February 26 read: "While I was making these observations, at 7:15, Rome time, Fr. Secchi died. Thus has Italy been deprived of its principal and most distinguished astronomer."

Secchi had shown the Vatican that having a presence in the world of science was a great way of being recognized as a nation independent of other nations. The pope who followed Pius IX, Pope Leo XIII, no doubt had this in mind when he established the modern version of the Vatican Observatory. One of its first activities was to take part in an international program to photograph and measure the entire sky, the Carte du Ciel. Eighteen observatories were given their own area of the sky to photograph. Italy had one swath; the Vatican had another.

Of course, it's ironic that the first activity of the observatory inspired by Secchi was to make the sorts of measurements—astrometry—that had been superseded by Secchi's work in astrophysics.

One can see in Angelo Secchi's career an illustration of the fundamental change in the nature of science itself during the quarter century of Secchi's greatest activity. For all that we honor in Galileo and Newton as the founders of modern science, science as we know it, for better or worse, is a nineteenth-century invention. And Secchi was a key figure in how that came about.

Secchi altered the nature of the questions that science would ask. Instead of "How are things arranged?" they asked, "What are things made of and how do they work?" And Secchi also altered the way that he went about that question in his work.

> For all that we honor in Galileo and Newton as the founders of modern science, science as we know it, for better or worse, is a nineteenth-century invention.

He embraced the new technologies of electricity and steam to change the way that data were collected. He pioneered the use of telegraphy to produce synoptic weather maps; and with newly devised batteries, wires, and electrical relays, he created his famous device to automatically record climate and weather data. At the same time, he took advantage of the ease of travel (and postal delivery) that steam-powered ships and trains provided to correspond regularly with scientists in Sicily, Paris, and Berlin, and also to travel routinely across Europe to observe eclipses and attend scientific congresses. This began with his own journeys to England and the United States in 1848–1850, which transformed his own understanding of the scientific enterprise.

Finally, and perhaps most radically, he found ways to integrate his own scientific interests with the needs of his national government (the Holy See), providing it with the expertise to deal with the challenges of the times. This included the response of the government to the growth of urban areas, with concomitant issues of public health, and the systemization of cartography and timekeeping. He also demonstrated to his government, and to the Catholic Church, how important and useful it was to have a visible and credible presence in the world of science.

The natural philosopher of 1801—toiling before even the word *scientist* had been coined—would not recognize the life of a scientist in 1860. In contrast, Secchi's day-to-day activities are easily recognizable to a scientist of today. From his time forward, science would be the enterprise of collaborations tied together by modern communication technology. The collection of data would depend more and more on complex instruments of a sort that required a team of supporting engineers. And support for science would be closely tied to government funding of both universities and national observatories.

Because Secchi's work was so wide ranging, we can find reflected in his life so many of the changes that stormed the intellectual world of the mid-nineteenth century. In them, we recognize the roots of contemporary science.

The curator at Stonyhurst College, Jan Graffius, has suggested that this nineteenth-century chasuble, which Hopkins would have seen, was an inspiration for his description of the "ah, bright wings" of the Holy Spirit.

CHAPTER NINE

THE POET AND THE COMET HUNTER

God's Grandeur

The world is charged with the grandeur of God.
   It will flame out, like shining from shook foil;
   It gathers to a greatness, like the ooze of oil
Crushed. Why do men then now not reck his rod?
Generations have trod, have trod, have trod;
   And all is seared with trade; bleared, smeared with toil;
   And wears man's smudge and shares man's smell: the soil
Is bare now, nor can foot feel, being shod.

And for all this, nature is never spent;
   There lives the dearest freshness deep down things;
And though the last lights off the black West went
   Oh, morning, at the brown brink eastward, springs—
Because the Holy Ghost over the bent
   World broods with warm breast and with ah! bright wings.

—GERARD MANLEY HOPKINS, SJ

YOU MAY RECOGNIZE THAT I QUOTED from this poem by Gerard Manley Hopkins, SJ, a few chapters back. "God's Grandeur" is perhaps his most famous work, for good reason. It shows both his unusual rhythms, his mastery of words, his deeply religious sensibility, and his common themes of love of nature and faith in the face of darkness. And it's a lot easier to understand and follow than many of his other poems!

When it comes to poetry and Jesuits, you can't do better than Hopkins. Even his personal story is the stuff of poetry. Born in 1844 in comfortable middle-class circumstances to a devout Anglican family, and the oldest of nine, he went on to attend Balliol College, Oxford, from 1863 to 1867. There he wrote poetry and made friends with Robert Bridges (1844–1930), who would become poet laureate in 1913. He was well on his way to a comfortable, conventional Victorian life. But while at Oxford he was attracted to the Oxford Movement of John Henry Newman, and in 1866 Newman accepted him into the Roman Catholic Church—not a popular choice with Hopkins's family.

In 1868 he resolved to enter the Jesuit order and burned all his poems, believing that was necessary in order to devote himself to God. However, in 1875, at the urging of his religious superiors he wrote "The Wreck of the Deutschland," about the tragic sinking of a ship. Although this marked the renewal of his poetic writing, the poem was not published in his lifetime, nor were most of his other poems. He spent his last years teaching at the Jesuit school in Dublin, far from friends and essentially unknown. He died of typhoid in 1889, at the age of forty-four. Only in 1918 did his friend Bridges bring out a volume of Hopkins's poetry. But by the mid-twentieth century—long after his death—he was recognized as one of the giants of nineteenth-century English poetry, with a style that was a century ahead of its time.

An interesting aspect of Hopkins can be seen in the poem that opens this chapter. Much of Hopkins's work pays close attention to the rhythms of nature, but in "God's Grandeur" we see a specific reference to the sky itself. In contrast to the soil, spoiled by man's toil, we look up with hope to see in the sky "the last lights off the black West" at sunset and "the brown brink eastward" of daybreak.

More than the sky, however, a close reading of Hopkins's poems, letters, and other documents shows that he was in fact a talented amateur astronomer. For David Levy, in his book *The Starlight Night: The Sky in the Writings of Shakespeare, Tennyson, and Hopkins,* "when Hopkins writes about the stars he discusses them as though he knows what he is writing about, and is comfortable doing it. . . . Instead of the generalized and vaguely resolved thoughts that comprise poems like Wordsworth's 'Star Gazers,' we find in Hopkins's astronomical references a conciseness and precision that approaches that needed for formal astronomical observations."

Who is David Levy to comment about Hopkins as both an astronomer and poet? Levy is a comet hunter, perhaps most famous for his co-discovery of Comet Shoemaker-Levy 9, which spectacularly struck the atmosphere of Jupiter in 1994. But in fact, Levy's academic career is not in astronomy but in English literature.

Levy lives near Tucson, not so far from our offices of the Vatican Observatory Research Group at the University of Arizona, and I have been proud to call him a friend for many years. In the spring of 2023, I visited him and interviewed him for this book.

AN INTERVIEW WITH ASTRONOMER DAVID LEVY

Tell me a little bit about yourself. Where are you from? How did you get into astronomy?
I was born in 1948 in Montreal; my dad was also from Montreal, my mother was born in New Orleans.

When I was young, I went to a summer camp. It was the Fourth of July . . . and as we're walking up the hill towards our cabin, I just happened to look up toward the darkening sky, and I saw a shooting star. And I looked, I looked at the others, and I said, "Did any of you see that shooting star just now?" And the other kids said, "No, we didn't." And then the thought entered my little eight-year-old David brain . . .: Was that message just for me? And so I put it in my brain and I left it there.

A few years later, I was in a bicycle accident and broke my arm. My cousin gave me a book about astronomy: *Our Sun and the Worlds Around It*. I read that book cover to cover and again and again and again, and I decided, *I'm going to do astronomy*. Finally, one day my dad at dinner said: "Last week, all we talked about were those damned stars. This week we're going to talk about something different! Next week we'll go back to the stars." What he was really saying was, "David, please don't make astronomy the most important part of your life." And I said to myself, *OK, Dad, I won't. I'll make it the only part of it.*

Jumping forward a lot, in April 1976, I joined a group of Montreal astronomers watching the Lyrid meteor shower. We were all looking up. It was a clear night. We saw a few, not too many. But for the first time in my life, I thought, *I wonder how many other people have seen meteors on this night? Or are doing that tonight, or have done that in the past? And how many people who are not astronomers, like writers, poets, artists, have also seen these, have seen the night sky?* And suddenly that was my career: to relate the night sky to English poetry and prose.

I did my master's thesis at Queens [University in Ontario] on Gerard Manley Hopkins. I did my doctoral thesis at the Hebrew University on Shakespeare.

So how did you discover Hopkins?

I was having my first meeting with Norman MacKenzie, who was my [master's] supervisor, and who later became a dear, dear friend. . . . I told him about what had happened just the very night before, about my thought about the people, English poets, seeing the night sky. And he said, "Well, then I must introduce you to Gerard Manley Hopkins. He was a Jesuit, but he loved the night sky."

Did you have any idea what a Jesuit was at that point?

A little bit, not as much as now.

What poetry did you start looking at? What poems in particular?
Most of us who have studied Hopkins think he's the most difficult poet in the universe to study. He uses sprung rhythm. It's a very complex rhythm where the feet are exchanged; it's almost like syncopated jazz. I mean the guy was so incredibly modern—fifty years, one hundred years ahead of his time.

He's extremely difficult to analyze, and it's very difficult to read his poems. And most students when they read Hopkins are saying, "When do we get on to some other poem? Any other poem." . . .

Except for this poem that he wrote—he was a student at Oxford before he became a Jesuit. . . . And he wrote a poem about a comet, and I'm going to recite it to you right now. Because I did memorize it. And when I read the poem, I was just blown away by it:

> —I am like a slip of comet,
> Scarce worth discovery, in some corner seen
> Bridging the slender difference of two stars,
> Come out of space, or suddenly engender'd
> By heady elements, for no man knows:
> But when she sights the Sun she grows and sizes
> And spins her skirts out, while her central star
> Shakes its cocooning mists; and so she comes
> To fields of light; millions of travelling rays
> Pierce her; she hangs upon the flame-cased sun,
> And sucks the light as full as Gideon's fleece:
> But then her tether calls her; she falls off,
> And as she dwindles shreds her smock of gold

Amidst the sistering planets, till she comes
To single Saturn, last and solitary;
And then goes out into the cavernous dark.
So I go out: my little sweet is done:
I have drawn heat from this contagious Sun:
To not ungentle death now forth I run.

I thought, "This is the man I want to write a thesis on . . ."

The main chapter of my thesis is about that particular poem. And Dr. MacKenzie suggested that I try to use my own knowledge of astronomy to see which comet it was referring to. I did; I came back, and I said, "Could it have been Donati's comet in 1858?"

Comet Whipple-Fedtke-Tevzadze, photographed in 1943 at the Vatican Observatory in Castel Gandolfo.

He said, "No, he wrote the poem in 1864. He would have forgotten about Donati by then."

I said, "You're right."

Then there was the great comet of 1861, which Tennyson saw. And Dr. MacKenzie said the same thing. It's still too far in the past. There was Comet Swift-Tuttle in 1862—again, no. Then I came across this one in August 1864, Comet Temple-Respighi. It was very bright for just a few days as it came close to Earth and then faded. I went with this to Dr. MacKenzie, but he said, "No, because you can't be certain of it."

I said, "I can."

He said, "Oh, tell me."

I said, "In the *Illustrated London News,* a very popular English paper at the time, there was a letter announcing the comet and they said, 'On Monday next it will be between the star Iota Aurigae, and Beta Tauri.' It was close between those two stars, so they were very easy to see. So that kind of shows that that was the comet he was writing about."

Then he said, "No, it doesn't. Because if it's between Iota Aurigae and Beta Tauri, Iota would be much, much fainter than Beta." [Stars in a constellation take Greek letters, normally from brightest to dimmest; so Beta Tauri would be the second brightest star of the constellation Taurus, while Iota Aurigae would be the ninth brightest star in Auriga.]

And I had a lot of respect for that man. "But you're wrong in this case," I told him.

And he said, "Well, tell me how."

And I said, "Auriga has many, many more bright stars than Taurus does. It turns out that Iota Aurigae is less than a magnitude fainter than Beta Tauri."

He said, "Are you sure?"

I said, "I'm quite sure. It had to be that comet. It couldn't be any other comet."

And right away he said, "Then that's going to be the main chapter in your thesis."

What does it mean to you as a "comet person" to be able to say he's not just writing about comets but that it is one particular, real comet?

My interest in comets has to do with every aspect of them—their history, their mythology, how they're referred to in works of literature, and then of actually observing them. I don't believe that particular comet [Comet Temple-Respighi] is periodic, so it would never come back, so I'll never have a chance to see it, but I feel that I know it very well because of its relation to Hopkins.

So the fact that it's a real comet and not just an imagined comet, does that make the poem stronger? Does it make it more real? Does it limit what the poet could say about it?
I think it makes it more real. The first thing you say, this comet was something that he saw and it touched his heart. And it touched his heart enough that he wrote about it. And over a century later, maybe a century and a quarter later, I got interested in it. He's a Victorian poet, and he's been gone for so, so long that it's almost like no one will ever see it; but in a way I feel that I have.

It's of course one of the astonishing things about Hopkins, among others, that he was completely unknown in his lifetime, and even for fifty years after his death.

Did Hopkins's style of poetry make this comet come alive in the poem the way that another style, another poet, another Victorian might not have been able to do?
The main way I can answer that question is to say that it was his style. It was his personality, his passion. Now his style of sprung rhythm and the details of it did not emerge until many years after he wrote that poem. So this is a very early poem, one of his first, and it is wonderful.

I remember when I got an honorary degree at Queens after Shoemaker-Levy 9, and they had kind of a celebration at the principal's house, and he said, "You wouldn't be able to tell them to read that poem to us, would you?" I said no, but I can recite it. And I remember Dr. MacKenzie was there with a big, big smile on his face.

Of course, Hopkins was a Jesuit; he was a deeply religious man. But this was written before he was a Jesuit. So . . . do you find his religious sensibility in the things he talks about being evoked in him as he sees the comet?

Oh yes. Especially his "Sonnets of Desolation" near the end of his life: "pitched past pitch of grief." And he really couldn't explain why he was in such a depression. And since I have a tendency towards depression as well, I really could understand that. And the "Sonnets of Desolation" are some of the most beautiful poems the man ever wrote. . . .

And, of course, he never recovered from that particular depression. I think he died when he was in his forties. I wish he were still around, sitting right there . . .

When you were first reading Hopkins, did you understand his religious impulses?
I did not understand them as I do now. I still don't understand them entirely. I don't think anybody understands anyone else's impulses. But I do enjoy, I do understand the passion; that I do understand. I understand that if he were sitting right here, he'd be asking all kinds of questions about Shoemaker-Levy 9!

Are there any other starry poems that Hopkins wrote?
He wrote "The Starlight Night," which was after he became a Jesuit:

> Look at the stars! look, look up at the skies!
>   O look at all the fire-folk sitting in the air!
>   The bright boroughs, the circle-citadels there!
> Down in dim woods the diamond delves! the elves'-eyes!
> The grey lawns cold where gold, where quickgold lies!
>   Wind-beat whitebeam! airy abeles set on a flare!
>   Flake-doves sent floating forth at a farmyard scare!
> Ah well! it is all a purchase, all is a prize.

Buy then! bid then!—What?—Prayer, patience, alms, vows.
Look, look: a May-mess, like on orchard boughs!
    Look! March-bloom, like on mealed-with-yellow sallows!
These are indeed the barn; withindoors house
The shocks. This piece-bright paling shuts the spouse
    Christ home, Christ and his mother and all his hallows.

. . . "Look up at the skies, look at the stars." And in the next line ["the circle-citadels"] he talks about the Corona Borealis. Well, nobody knows that for sure, I'm sorry, except him. But I'm probably right.

But it was the growth of maturity between the comet poem and this one that was just incredible to me. He was now not just thinking as someone who lives under the night sky, as he did in his youth. But it was someone who was waiting for the night sky to show his fate.

How old was he when he wrote this? Where was he?
He was probably in his early thirties. He had had his "Slaughter of the Innocents," as he called them. When he became a Jesuit, the rule at the time was that [they were] not supposed to write any poetry.

But anyway, in these later years he was seeing the passion that the stars aren't just points of light in the sky of great distances. They are like a celestial tent. . . . King David somewhere in the psalms talks about a "celestial tent for me to dwell in," which is one of the quotes that I use when I'm giving a lecture sometimes. And I enjoy the thought of that.

The Jesuit school of Stonyhurst, Lancashire, England, where Gerard Manley Hopkins, SJ, taught as a young Jesuit; photographed from the site of its astronomical observatory where Secchi once observed.

Sunset at St. Beuno's, in Wales, the Jesuit retreat house where Gerard Manley Hopkins, SJ, wrote "The Wreck of the Deutschland."

But I think the main thing is that it isn't just a group of facts related to a group of stars. It is the entire picture of the night sky.

We're warned many times in Scripture not to worship the night sky, but it doesn't say anything about not loving it and going out and spending our time loving the night sky, as I think you're aware as well.

There is a real difficulty for anyone who looks at the sky to get past the "it's all one dome, it's all just a bunch of stars at the same distance" before realizing that we are in a three-dimensional void. The first time that I recognized that difference was when I was in the Southern Hemisphere and saw the Magellanic Clouds. And that made me realize, oh, everything else is the Milky Way, and I'm in the middle of that, but I'm not in the middle of those clouds. Do you think Hopkins was beginning to see the sky in a depth both metaphorical and real?

Oh, I think so. I think he was trying to understand the cosmic picture of the night sky. As I like to say, when people move to Arizona, most people move here because of the environment. That environment includes the night sky. . . . And I think Hopkins would have felt that way as well. I kind of imagine him nodding his head there and saying, "You guys don't have any idea what you're talking about. Right, right." What if you could put that into a description? You just told this marvelous story of G-d as somebody learning how to write a modal exercise.

Who do you think God was to Hopkins?

I have to begin answering that by saying I don't know—of course, of course. I think that G-d was an inspiration, and I think Hopkins would agree with this. . . . In fact, in his first few years as a Jesuit there was a terrible accident at sea and this boat sank at five miles from shore. And he was advised by his superior to resume his poem writing and to write one about the "Wreck of the Deutschland." And that was his first major poem. And I think after that, they said, "Why don't you bring back your earlier poems and write poems?" He could write poems now, and he did.

One last thing. Is there anything about Hopkins's understanding of the sky that you would say was particularly Jesuit?

In the sense that the Jesuits prize scholarship above almost everything else . . . yes. I think that just to go out and look at the sky and feel it and see it and becoming blind still staring at the sky is a good part of it. But if you have no idea how far a star is, how many galaxies there are, and the other thing, you know, if you have no idea, then I think you're missing more than two-thirds of the story. And I think it's worth it to learn about what's happening.

You may have heard this story . . . when the James Webb Space Telescope was launched, and Bill Nelson, the administrator of NASA, was giving a lecture the day of the successful launching. And he was giving a list of some of the people who helped. And it was kind of boring. And then he said, "And then there's one other person who lived thousands of years ago . . . a shepherd boy named David; and he sat down, and he looked up at the stars, and he wrote poetry."

Oh, that got to me. Oh my G-d, yes.

We're warned many times in
Scripture not to worship the night sky,
but it doesn't say anything about not
loving it and going out and spending
our time loving the night sky.

Sunrise across the Mississippi River, as seen from the Jesuit White House Retreat Center in Oakville, Missouri

CHAPTER TEN
JESUIT COSMOLOGIES

WHEN ONE THINKS OF CREATION IN SCRIPTURE, the opening of Genesis immediately comes to mind: "Let there be light!" There's a great temptation these days, contemplating the Big Bang and the creation of the universe in a burst of energy, to look at the opening verses of Genesis and smirk—*Hah! We got it right!*

Wrong.

If there is a theme to a modern Jesuit understanding of cosmology, it is precisely in rejecting that easy match. The belief that you can ferret out in Scripture cleverly disguised versions of the latest discoveries of science is called "concordism," and it's a seductive temptation.

Pope Pius XII, in an address to the Pontifical Academy of Sciences in 1951, came dangerously close to endorsing the Big Bang on the principle that it reflected the beginning point described in Genesis 1. (It took an intervention by Fr. Lemaître himself, the author of the Big Bang theory, to convince the pope *not* to endorse his theory.) We've honored the pioneering science of Angelo Secchi, SJ, whose understanding of stellar spectroscopy ultimately gave us the tool needed to measure the expansion rate of the universe, but his popular books on science and faith were riddled with concordism. Even Galileo himself, in a famous letter to Christina of Lorraine, Grand Duchess of Tuscany, fell into the trap of trying to defend his science by drawing parallels to what can be found in Scripture.

But Pope John Paul II, in his letter to George Coyne, SJ, the Jesuit head of the Vatican Observatory, came up with the most succinct reason to reject concordism: "Both religion and science must preserve their autonomy and their distinctiveness. Religion is not founded on science nor is science an extension of religion."

Concordism tries to force religion and science together in a way that does not preserve their autonomies or their distinctiveness. Worse, the desire to merge them comes from a false idea of our ultimate goal.

Too often science and religion are presented as rivals for our affection, competing gods, each demanding total worship. Concordism forgets that neither science nor religion is an end in itself. "Faith and reason are like two wings on which the human spirit rises to the contemplation of truth," wrote Pope John Paul II in his encyclical letter *Fides et ratio*. Truth is the goal; science and religion are just ways to help us get there.

Besides being bad theology, concordism has another obvious problem. What happens when time passes on, and this year's latest insight becomes next year's exploded theory? For example, the science in Genesis is likely 2,600 years old. It was the "best science" of its time, but it is obsolete today—indeed, it was obsolete by the time of Aristotle. So why do we continue to read Genesis? Because Genesis is not a book about creation, about science, or rocks or planets or stars. It is a book about God as Creator. We are told in Genesis not *what* God creates, but *how* God creates.

Genesis was the product of Jewish scholars in Babylon during the Babylonian captivity, the sixth century before Christ. As Pope Francis notes in *Laudato si´*: "The experience of the Babylonian captivity provoked a spiritual crisis which led to deeper faith in God. Now His creative omnipotence was given pride of place in order to exhort the people to regain their hope in the midst of their wretched predicament."

The Genesis version of creation accepts the arrangement of the heavens and earth assumed by the Babylonians circa 600 BCE: a flat disk with a vault overhead separating the waters below and the waters above. And there are many elements of the creation story that mimic Babylon's creation myth, the *Enuma Elish*. But it is not how Genesis follows the Babylonian creation story that matters; what is important is how it is different.

In the Babylonian story, the universe was the result of a fight between the god Marduk and the dragon of watery chaos, Tiamat. (In the opening verses of Genesis, the name of the dragon, Tiamat, is turned into the Hebrew word used for "watery chaos.") The dragon's corpse became the physical universe. And to the Babylonians the ultimate pinnacle of creation was the formation of the city of Babylon, gateway to the gods.

Why did the authors of Genesis want to write their own creation story? What were they trying to say? How did it go beyond what the Babylonians had already said? Central to understanding the Genesis creation story is that it is written in reaction to that story, against the surrounding culture of Babylon.

First of all, in Genesis, the Creator is one, not many. And the creator God is already present when the story begins "in the beginning." This immediately removes the God of Abraham from any of the pantheons of nature gods found in the surrounding cultures. God is not a nature god. Rabbi Jonathan Sacks, in his book *The Great Partnership,* calls this "the discovery of God beyond the universe." God is outside of nature, already present in the beginning. God is supernatural. And only something outside of nature can give meaning to nature.

This Creator does not serve as a replacement for the laws and forces of nature. The God of Genesis is not someone who throws thunderbolts at whim. Rather, this Creator becomes the One who allows for no nature gods, and thus makes room for a universe that follows laws, a God who accounts for the existence of such laws and forces in the first place.

It also means that we can learn those laws, and thus get to know the personality of the Creator in the timeless, elegant, and beautiful character of the laws that underlie nature—laws that give the universe the freedom to act in ways that are good or ways that

Pope Francis and the author.

are not good. And as Pope Benedict pointed out, creation occurs in a way that everybody can see. When God says, "Let there be light," it means that nothing is hidden; nothing is stopping us from discovering those laws.

Creation occurs through the deliberate decision of the one God, present before creation, acting alone. Pope Francis quotes Psalm 33 in *Laudato si´*: "'By the word of the Lord the heavens were made.' This tells us that the world

> We are called to see, to love, and then to understand.

came about as the result of a decision, not from chaos or chance, and that exalts it even more. The creating word expresses a free choice. The universe did not emerge as the result of arbitrary omnipotence, a show of force, or a desire for self-assertion."

The creation of the universe is described in Genesis as occurring step-by-step, in a sequence as logical as day follows night. Rather than invoking nature spirits to make things happen, the universe is allowed to operate by its own internal rhythms. That is, God chose to make science possible.

And even more striking, we are told at every step that creation is good. (Another scholar has informed me that the Hebrew word we translate as "good" could also mean something to the effect of . . . "yeah, that works." Which is of course an engineer's highest praise!) The Creator delights in this physical universe and the way it works. The physical universe is not something dangerous, inherently evil, and sinful; nor should we reject our created nature and try to just be "spiritual."

Finally, what is the point of creation? Why was the universe made? While the Babylonian creation myth worked to promote the city of Babylon as the logical climax of creation, Genesis has a very different destination. The ultimate end of creation is the seventh day, the day of rest: "So God blessed the seventh day and hallowed it, because on it God rested from all the work that he had done in creation" (Genesis 2:3).

The Sabbath is not just a day of idleness. It is a day of prayer, for the contemplation of God and the works of God. The other six days cover the work we must do to survive: food, shelter, and clothing. And these are good. But contemplating creation as a way of engaging in the Creator is different. It is a form of prayer.

The human being, also a creation of the Creator, is more than an animal who lives only to eat and sleep. The spiritual and intellectual aspects of the human person are what makes us more than other animals. Our ultimate call is to praise God in the universe, to find God in all things. We are called to see, to love, and then to understand.

We were created to praise the Creator. How do we do that? Each in our way: to observe and love what we see, to write poems or scientific treatises, and to share with one another what we have seen and experienced. We were created to be astronomers!

St. Paul reminds us in Romans 8:19 that "creation waits with eager longing for the revealing of the children of God." Creation could not be revealed until we were present and ready to receive that revelation. We become the children of God when we become aware of God's creation and thus, in it, of God's presence.

One way that happens is when we appreciate the stars. On a large plaque attached to the building housing one of our vintage telescopes in Castel Gandolfo is the motto given to the Vatican Observatory by Pope Pius XI in 1935: *Venite adoremus, Deum creatorem,* or "Come, let us adore God the creator." (See photo on page 163.) We, ourselves, creatures of that very same Creator, have become the consciousness of creation. In us, creation finds God. And when we find God, how can we not give praise?

But even if contemplating the universe with the eyes of a scientist might be encouraged by the Genesis story, can we really say that doing so is *necessary*? We live our daily lives without ever thinking about the fact that we're walking on the surface of a sphere; much less, that this globe that seems so big to us is an insignificant speck orbiting an average star in an average galaxy, one of billions of such stars and billions of such galaxies.

Pope Pius XI at the dedication of the new telescopes at Castel Gandolfo.

And so one might ask, do we need to know about all those galaxies? There were saints long before there was modern cosmology. The writers of Genesis, who described the universe as a disk covered by a dome, knew nothing of modern astrophysics. But they knew God well. Isn't living a good life good enough for us?

In answer to this, the book of Job argues otherwise. When Job, in his crisis, complains to God about the injustices he has endured, God responds, in a fascinating way, by insisting that to be self-satisfied with our ignorance is to limit both ourselves and our relationship to God.

We have seen how Job uses the immensity of creation to express despair in ever being able to understand or face God: "'How can a mortal be just before God? If one wished to contend with him, one could not answer him once in a thousand'" (Job 9:2–3). And yet it is clear that God does not want Job, or us, to despair. Later, God challenges Job to confront how little he knows of all the wonders of the universe. These include not only the wonders of what can be seen, like the constellations, but also the reasons that explain how those things came to be, things that Job had never thought about before. Recall these verses from Job that we first encountered at the beginning of this book:

> Then the LORD answered Job out of the whirlwind:
> 'Who is this that darkens counsel by words without knowledge? . . .
> I will question you, and you shall declare to me.
>
> 'Where were you when I laid the foundation of the earth?
> Tell me, if you have understanding.
> Who determined its measurements—surely you know!
> Or who stretched the line upon it?
> On what were its bases sunk,
> or who laid its cornerstone
> when the morning stars sang together
> and all the heavenly beings shouted for joy?'
>
> JOB 38:1-7

The message we had highlighted at the beginning of the book was how the stars acknowledged God as Creator. But now there is additional significance to us, observing those stars. Just a few verses later, God challenges Job: "'Where is the way to the dwelling of light?'" (Job 38:19) and "'Do you know the ordinances of the heavens?'" (Job 38:33).

Why does God ask this? To show off his greatness? To intimidate Job? To put Job's problems in perspective? No—an invitation is implied. Come with God, and discover the way to the dwelling of light. We are invited to join with the morning stars who sang in chorus while all the children of God shouted for joy at the foundation of Earth. Only by doing so can we—and Job—be in a position to fully appreciate who God is and how he expresses his love.

Anyone can take part in the game of being fascinated by the workings of the universe, at least as a spectator. But for those of us who are called to be astronomers, the exploration of God's creation is a response to a special invitation to spend time with the Creator. We get to play with him, so to speak, uncovering the delightful puzzles he sets for us and marveling at the way the laws of the universe fit together with a logic that is both harmonious and elegant. In this way, we learn to see a side of God's personality.

There are many other places in Scripture where God's role in creation is discussed in detail, but I conclude here by noting one final curious story. Gottfried Wilhelm Leibniz (1646–1716) famously asked in his essay "On the Ultimate Origin of Things," "Why is there something instead of nothing?" Creation could not have created itself. Why

> Creation could not have created itself. Why does creation itself exist?

does creation itself exist? If we say it was created from something preexisting, whether the primordial chaos of Genesis or the space-time continuum of modern cosmology, we still have to ask where that preexisting chaos or space and time and laws of nature come from.

In the second book of Maccabees, however, we find a surprising answer: God created from nothing, *ex nihilo*. And the real surprise comes from the fact that this idea of "creation out of nothing" is tossed off almost in passing, as a commonplace. When King Antiochus tortures seven sons, their mother encourages her youngest to remain firm in his faith by reminding him to "look at the heaven and the earth and see everything that is in them, and recognize that God did not make them out of things that existed" (2 Maccabees 7:28). The mother understands not only the order of chaos but also that the chaos itself needed to have been created by a supernatural God, outside of space and time.

William R. Stoeger, SJ (1943–2014), a cosmologist at the Vatican Observatory, wrote extensively about the difference between the scientific and philosophical questions of origin. He emphasized the difference between asking, "Why is there something instead of nothing?" and "How did things get started?" The Big Bang, even though it was first suggested by a Catholic priest, Georges Lemaître, is not the same thing as the Genesis creation, nor is it the same thing as "creation out of nothing." God's creation includes the creation of space and time, and thus from our vantage point, it occurs at every space and every time. This is in accord with the Thomistic insight that God does not merely start the universe like a philosopher's "prime mover" but continually maintains its existence.

In creation are made the laws of nature themselves, along with the fabric of spacetime that they can operate upon. Creation of time by the supernatural Creator must occur outside of time, and thus at every time, maintaining the universe at every place and every time. The Big Bang is just the story our current science tells of how it unfolded.

Looking at the stars with the naked eye, we can see two obvious things. First, the stars are not evenly distributed in the sky. And second, the stars are visible in a sky that is dark at night. If we lived in a simple, static, infinite, and unchanging universe, neither fact would be true!

Let's assume that the Sun is a star (which it is) and that most stars are like the Sun (this is not actually correct, but for our purposes, the difference doesn't matter). Now let's assume that the universe is full of those stars, randomly distributed, from here to infinity.

The farther away any given star is, the dimmer it will appear. But the farther away we look, the more stars we will see. It turns out, these two effects exactly cancel each other out. (For those of you who

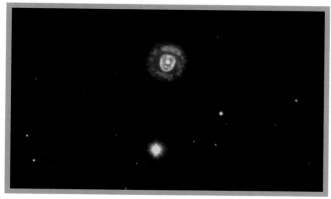

This marvelously odd looking planetary nebula is sometimes whimsically called "Clown Face." It is actually a ball of gas erupted from a dying star, located in the direction of the constellation Gemini; it lies about 6,500 light-years away; Vatican Advanced Technology Telescope image.

are interested: the light from a star at a given distance, R, drops as 1 over R^2, while the number of stars at a given distance from us, R, grows by R^2.) Another way to think of this is that if the universe actually were infinite and full of stars, then no matter which direction we look, eventually our line of sight would encounter a star. The result? The sky should be white with starlight. But it isn't. Why?

In the textbooks this is called Olbers's Paradox, named for the German astronomer Heinrich Wilhelm Olbers (1758–1840), who, of course, was by no means the first person to think of it. Indeed, Kepler had an inkling of this problem around the same time as Galileo. Kepler's answer was clever and wrong—he decided that the stars were not suns.

For various reasons, which seemed reasonable at the time but turned out to be fundamentally wrong, he concluded that stars were actually much larger than our Sun but much dimmer.

Part of the answer to Olbers's Paradox is that there's also dust in the universe, sometimes in thick clouds but also spread through the empty space between the stars. That dust has an effect in dimming the light from more distant stars. But there's a more important reason the sky is not white with light even at night: the simple fact that the assumptions behind Olbers's Paradox aren't true. The universe we see is not infinite, and what we see of it is not uniform.

All the stars that we can see with our naked eye, even on the darkest night in the most wonderfully dark site, are set inside a large but finite disk of stars: our own galaxy. It's a disk that looks pretty much like other galaxies we see in our telescopes. Most of the time when we are looking at the night sky we're looking up (or down) out of the disk with just a relatively few stars in our field of view and nothing but dark empty space (and distant galaxies) behind those stars. There's not an infinite number of stars in those directions.

But looking in certain directions in the sky, where the stars are richest, we eventually find ourselves looking lengthways into the disk itself. And in those directions, we do indeed see that the sky turns white, a path of light across the black sky. We call this the Milky Way. You need a very dark sky to see the Milky Way in its full glory—it is just one of the marvels of the night sky that light pollution has robbed from our eyes.

So, our galaxy—the word *galaxy* comes from the Greek word for "milk," by the way—is finite. The only place it is thick enough to make Olbers's Paradox even a little visible is the Milky Way.

But what if there are an infinite number of other galaxies out there? The same paradox should hold, shouldn't it? Shouldn't the sky be filled with all the other Milky Ways out there? There are two reasons that doesn't happen, and both are explained by the Big Bang theory.

First, the universe itself isn't infinite, or at least, what we can see of it is not infinite. Space is not static, but it grows. The space between clusters of galaxies is constantly getting bigger, which means that the galaxies appear to be moving away from us, and the farther they are, the faster they appear to move, because there is more of that expanding space between them and us. Eventually if we are looking for galaxies beyond a certain distance—slightly closer than fourteen billion light-years—their motion away from us appears to be faster than the speed of light, so the light they emit is carried away from us faster than it can reach us. We can never see it. So, there's a limit to how far we can see.

In addition, because of that motion, the color of the light itself shifts from the kind of visible light our eyes might be able to see to red light, then infrared light, and beyond, wavelengths so stretched out in the expansion of space that they are no longer visible to our eyes, and no longer able to add to skyglow.

As it happens, the Big Bang theory also explains a lot more than that. It fits in with many other subtle facts about the universe. This is why it's the basic paradigm of all our cosmology. But the darkness of the night sky is the sort of thing that Olbers recognized as a problem, and that makes us realize that we need to think about a universe that's more complicated than just a static, infinite jumble of stars.

We know from experience that this is not the end of the story. There's so much more to learn, to understand, to correct, to get right, to let ourselves be amazed by. What we call the Big Bang today is not just the idle speculation of a few dreamy astronomers, but a complex, detailed mathematical construction based on Einstein's theory of general relativity. Of course, just because it is complicated and mathematical doesn't mean it's correct! Indeed, the fact that we are still working on it shows that those who know it best also are best aware of its shortcomings.

William Stoeger, SJ, spent much of his career working out ways to tie various possible theoretical descriptions of the Big Bang to the kinds of observations of galaxies that could be made at a telescope. Other Jesuits today, working at the Vatican Observatory and elsewhere, are pondering everything from the faint microwave radiation left over when the Big Bang occurred to developing the strange physics—quantum gravity—that might describe what occurred in the barest instants after the Big Bang itself.

Meanwhile, our understanding of the universe grows stranger and more marvelous. Dark energy and dark matter are the names we give to ingredients of the universe that we cannot directly detect but that we posit need to exist in order to explain the motions of the universe that we can directly detect. Is there a Jesuit connection to them? Well, one of the pioneers who studied the orbits of stars in other galaxies that strongly imply the existence of some sort of unseeable matter there was Vera Rubin (1928–2016), and she taught astronomy at Georgetown University, the Jesuit school in Washington, DC. Other players in the field, folks looking for evidence that would confirm or deny various proposed ways that this mysterious matter or energy might exist, are Jesuit colleagues, including folks who have used our telescopes and held workshops at our facilities in Rome and Arizona.

But it's stretching it to claim much of a Jesuit nature to this work. Rather, we Jesuits are working alongside a much larger community, worldwide, fascinated by these questions. In that sense, there is no more a Jesuit cosmology than there is a Jesuit moon.

Does unseen matter actually have any effect we can see for ourselves? Does it have any effect on the nature of our own existence? And what does all of this mean to the question of how well we can trust science?

Such questions weren't yet part of the curriculum when I was studying at MIT fifty years ago. They are examples of the fascinating things that astronomers will be looking for over the next hundred years. And yet they are but new wrinkles to age-old questions: How did the universe begin? How do planets form? What is life, where and how do we look for it, how do we know it when we see it, and what do we do then? Are there new laws of physics to be found in the extreme environments of astrophysics?

These are questions of origins. These are questions of our place in the universe. These are questions of knowing and telling. And all these questions remain even as the answers evolve. The questions matter more than the answers.

And what does all this mean to the rest of us? It means that so much more remains to be found. And that of course means that it's crazy to build your faith on today's science, which will be made obsolete by tomorrow's science. However, it is just as foolish not to let science enrich your faith with new reasons to be amazed. It means that the universe is stranger than we used to think but not stranger than we can think. It means that science is never at an end, but "the end of science" is not the end reason for which we do science.

And then we can open Genesis again and read the words of God the Creator with new eyes. The endurance of religion even after thousands of years of radical changes in our cosmologies reminds us of the amazing resilience of truth, even as the cultures that gave birth to that truth evolve beyond recognition.

Sunset as viewed by Spirit, one of the NASA rovers on Mars; NASA image.

CHAPTER ELEVEN

BE PRAISED, MY LORD, THROUGH THE STARS

POPE FRANCIS IS A JESUIT, and his work is infused with Ignatian spirituality. To understand what that means, just compare his 2018 encyclical *Laudato si´* to the original medieval poem "Canticle of Brother Sun and Sister Moon," written by St. Francis of Assisi, which begins with the words that inspired Pope Francis, "Laudato si´."

St. Francis of Assisi wrote his wonderful poem to evoke awareness of God in creation, but it is very different from how a Jesuit would express the same ideas. Assisi turns the Sun and stars into people—brothers and sisters—to show how they can be the subjects of our love. To a scientist, or a Jesuit spurred to find God "in all things," this anthropomorphism may sound odd. However, it is charming!

The phrase *Laudato si´* is medieval Italian for "Be praised." St. Francis's "Canticle of Brother Sun and Sister Moon" begins: "Be praised, my Lord, through all Your creatures: especially through my lord Brother Sun, who brings the day; and You give light through him. And he is beautiful and radiant in all his splendor! Of You, Most High, he bears the likeness."

Many images of the Sun come to my mind when I hear this description, but one that has a special flavor to me is a sunset as viewed by Spirit, one of the NASA rovers on Mars. Sunsets are so evocative in any setting; to see one on another planet makes the reality of that other place feel real.

The poem continues: "Be praised, my Lord, through Sister Moon."

And how many memorable images of the Moon there are from the Apollo era! Again, the ones that mean the most to me are the ones that remind me of the Moon as a place where people like you and me have walked. I actually got to meet the Apollo 17 astronaut Harrison Schmitt, the only trained geologist to go to the Moon. (See photo on page 175.) We've been to the same scientific meetings, and he once showed up to a bookstore in Albuquerque where I was signing one of my books!

It continues: "And through the stars; in the heavens You have made them bright, precious and beautiful."

Forty years ago, searching for a good dark site for a new set of telescopes, astronomers from the University of Arizona took an image of the full sky at the place that would become the Mount Graham International Observatory—home today of the Vatican Advanced Technology Telescope. (See photo on page 176.) The long exposure shows us sitting amid a disk of stars spreading out and around us from the concentrated light of the Milky Way, and following the Milky Way now hidden under our feet, to the other side of the world. The stars are not just something out there; we are amid them all.

"Be praised, my Lord, through Brother Fire, through whom You brighten the night. He is beautiful and cheerful, and powerful and strong."

The fire image that comes to my mind is not fire at all, but a large solar flare. Modern science knows that solar flares are hardly fire, but St. Francis wouldn't have known the difference.

Juan Casanovas, SJ (1929–2013), helped establish the first solar observatory on the Canary Islands in the 1960s, and then he came to work at the Vatican Observatory. Claudio Costa, the imaging specialist and engineer at the Vatican Observatory who has used our telescopes to produce so many of the great astronomy images in this book,

Geologist Harrison Schmitt standing next to a massive rock, surrounded by the
yellow soft mountains of Taurus-Littrow on the Moon; NASA image.

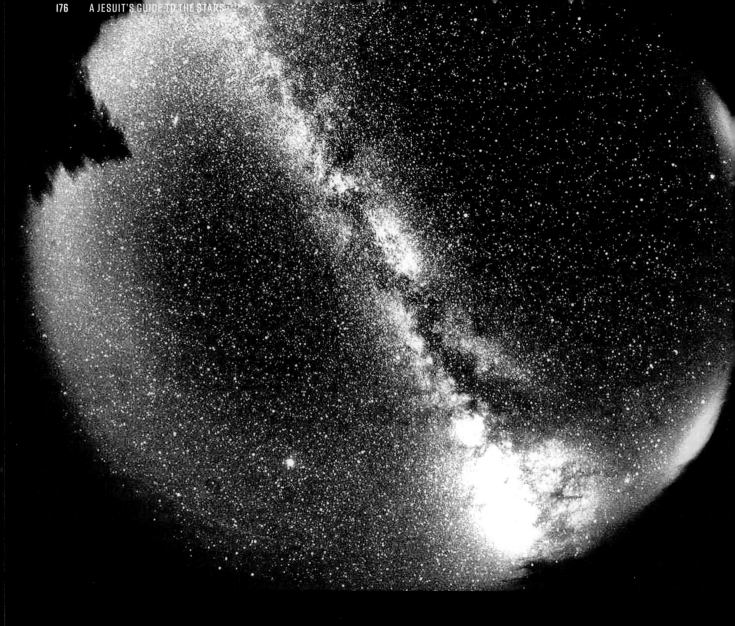

Long exposure photo of the full sky taken in the mid-1980s at the location that would become the Mount Graham

captured a burst of solar flame, a solar prominence, using the Coronado H-alpha solar telescope donated to the Vatican Observatory by its inventor, David Lunt (1942–2005), just before the transit of Venus in 2004.

Every time I see an image of a solar prominence, I think of Juan and Claudio and David.

"Be praised, my Lord, through Brothers Wind and Air, fair and stormy, all weather's moods, by which You cherish all that You have made."

I spent a season in Antarctica in late 1996, hunting meteorites on the Antarctic Plateau. For six weeks, half a dozen of us lived in tents not all that different from what the first Antarctic explorers used a hundred years earlier. Every morning, we headed out to search across the blue ice for our black meteorites; the meteorites just sat on the surface, and it was merely a matter of going out and collecting them. But one day it snowed, covering everything. For the next ten days we were trapped in our tents as the high winds blew the snow against the canvas, making a nonstop racket. Finally, the winds died down. By then, the snowfall covering the blue ice and meteorites had all been blown away, and we were able to go meteorite hunting again.

"Be praised, my Lord, through Sister Water; she is very useful, and humble, and precious, and pure," wrote St. Francis.

There are in fact whole worlds made of ice: the moons that orbit the giant planets of the outer solar system. Back when I wrote my master's thesis fifty years ago on how icy moons might melt in their interiors, I never expected to actually be able to see the water itself. But the Cassini-Huygens mission to Saturn took an image of water erupting out of the south pole of the icy moon Enceladus!

"Be praised, my Lord, through our Sister, Mother Earth, who sustains and governs us."

Can we see Mother Earth the way we see the other planets? Indeed, we can . . . if we go to those planets and look back. The Cassini mission to Saturn provided an excellent opportunity for just such a look. It was deliberately targeted to observe the rings of Saturn while the planet itself blocked the glare of the sun. Sunlight glowing through the rings lights them up like a beam of light in a dusty room. The bigger chunks of ice that make up the rings we see from Earth just block the sunlight and reflect it back to us, but small dust particles can scatter the sunlight sideways and forward. Thus, by moving past the rings and then looking back, we can discover rings of dust not visible from Earth. And there, in the same image, we also see Earth itself.

St. Francis's poem continues: "Praised be You, my Lord through Sister Death, from whom no-one living can escape. Woe to those who die in mortal sin! Blessed are they She finds doing Your Will. No second death can do them harm."

We speak of the life and death of stars, as if stars were alive. Whether or not that is accurate depends on what you mean by *life*. Certainly, stars do eventually run out of fuel, when all the lighter elements in their interiors have been completely fused into heavier elements and further fusion power, the energy that gives the stars light and heat, is no longer possible. When the center of a star can no longer be hot enough to hold up the massive weight of the rest of the star, the star collapses. Sometimes this leads to a puff of interstellar dust erupting from its surface, like the planetary nebulae we showed earlier. But if the star is large enough, this collapse produces a supernova, briefly shining as brightly as all the other stars in its galaxy. After it fades, it leaves an expanding cloud of gas.

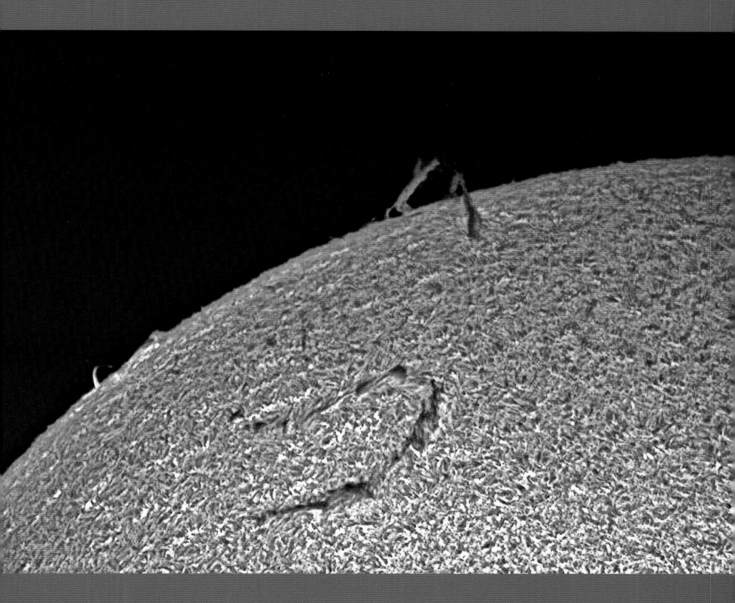

The burst of solar flame, a solar prominence, imaged by Claudio Costa using the Coronado H-alpha solar telescope donated to the Vatican Observatory by its inventor, David Lunt.

"Praise and bless my Lord and give Him thanks, and serve Him with great humility."

The spirituality of St. Francis called us to look and see all those creatures, all those bits of creation—Brother Sun, Sister Moon, and the rest, each called out by name, each individually giving praise to their Creator. Pope Francis looks at the same Earth and sees the same elements as St. Francis did, and indeed as St. Ignatius imagined the Trinity seeing. But Pope Francis adds yet a new perspective. He connects this gaze to the consequences of our society's injustices and our own sins.

The poisoning of Earth occurs ultimately because we are humans, partakers and subjects of sin. But as the first week of Ignatius's Spiritual Exercises teaches us, this is not a cause for despair. The scientist, the engineer, we users of technology—we have indeed all been sinners. But we are also loved by the One against whom we have sinned. Because of that love, we are also capable of human brilliance. We are capable of good. We all are creatures, created bits of this universe, all subject equally to its laws, which are God's laws. And while those laws are relentless if we ignore them, they can also bring out beautiful and even fun aspects of this universe. And what we have messed up, we can also clean up.

Snow blown by high winds across the ice plateau in Antarctica, where we searched for meteorites.

We must accept that we face an ecological crisis today. Technological fixes are important and necessary. But because the root problem is human sinfulness, not technology, then as long as we are blessed with free will, we will always have to face the curses of our choices. If nothing else, we know from experience that every technological fix inevitably comes with unintended consequences.

We will never come to the point at which we won't have to keep worrying about the care of Earth. We will never come to the point at which we won't have to worry about making mistakes. There is no system or technological solution that will relieve us of that continual responsibility.

> We will never come to the point at which we won't have to keep worrying about the care of Earth.

T. S. Eliot reflected on the nature of such misguided hope for quick fixes in his choruses from *The Rock,* a pageant play written in the 1930s, when all sorts of "isms" were competing with visions of a perfect world:

> They constantly try to escape
> From the darkness outside and within
> By dreaming of systems so perfect that no one will need to be good.
> But the man that is shall shadow
> The man that pretends to be.
>
> —T. S. ELIOT

There is no system so perfect that no one will need to be good. And we will never get it right all the time.

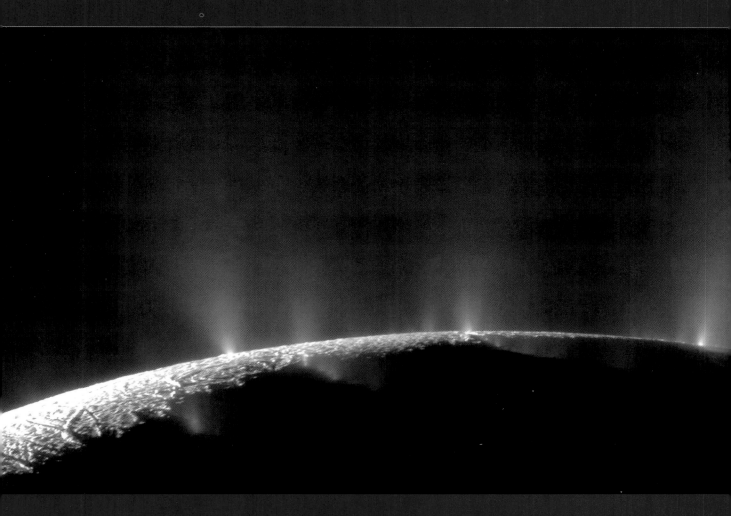

The Cassini-Huygens mission to Saturn took this image of water erupting out
of the south pole of Saturn's icy moon Enceladus; NASA image.

But that's no excuse to give up trying. Rather, it's a reason to stop beating up ourselves, or our neighbor, for not being perfect. Self-righteousness, thinking that we're better behaved than everyone else or smarter than everyone else, is never a particularly useful attitude. If God loves the sinner, can we love any less? Including the sinner that is ourselves? That, too, is a lesson of the Exercises.

This is one important difference of Pope Francis's *Laudato si'* compared to other popular reactions to the ecological crisis. Of course, we are not gods who can dominate Earth. But neither is nature a god, to whom we must bow and submit ourselves in such awe that we let it dominate us. In the words of the writer G. K. Chesterton, reflecting on St. Francis:

Saturn imaged by the Cassini mission. Earth is the dot at lower right; NASA image.

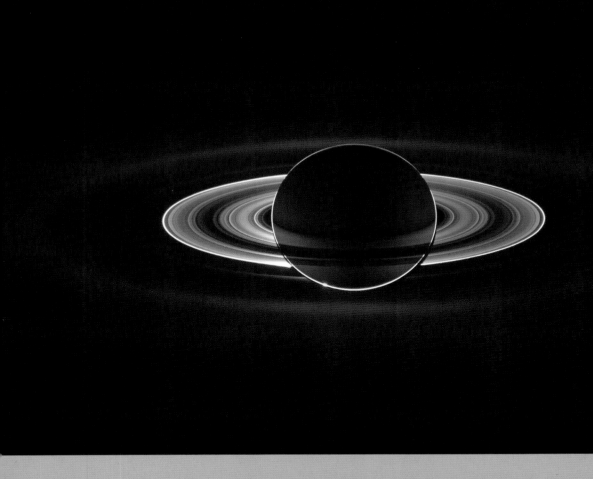

This Cassini spacecraft image was taken from the far side of planet Saturn opposite the Sun to highlight dust in its rings, including newly visible rings of only dust; NASA image.

The essence of all pantheism, evolutionism, and modern cosmic religion is really in this proposition: that Nature is our mother. Unfortunately, if you regard Nature as a mother, you discover that she is a stepmother. The main point of Christianity was this: that Nature is not our mother. Nature is our sister.

We can be proud of her beauty, since we have the same father; but she has no authority over us; we have to admire, but not to imitate. This gives to the typically Christian pleasure in this earth a strange touch of lightness that is almost frivolity.

Nature was a solemn mother to the worshippers of Isis and Cybele. Nature was a solemn mother to Wordsworth or to Emerson. But Nature is not solemn to Francis of Assisi or to George Herbert. To St. Francis, Nature is a sister, and even a younger sister: a little, dancing sister, to be laughed at as well as loved.

> The main point of Christianity was this: that Nature is not our mother. Nature is our sister.

You don't dominate your little sister. The very thought of doing so is disgusting, horrific. But you don't worship her, either. We, nature and humanity, are both children of the same Father. We interact with our siblings not out of fear or power, but out of love. And laughter.

The Crab Nebula, a remnant of a supernova, as imaged at the Vatican Advanced Technology Telescope by Br. Jonathan Stott, SJ.

Sunset, looking across Los Angeles to the Pacific Ocean from Mt. Wilson, near the hundred-inch telescope used by Edwin Hubble (1889-1953) to measure the expansion of the universe.

CONCLUSION: JESUIT ASTRONOMY

WHAT MAKES A SKY A JESUIT SKY? What makes stars Jesuit stars? From what we've seen here, and from my own experience as a Jesuit astronomer, I can see three distinctive traits of a Jesuit astronomy.

The first is the reason we do astronomy, study the stars, go out at night and look up at the sky. Obviously, it's not for profit or power or glory; there's none of that to be found in astronomy. Joy is what gets an astronomer up in the morning. What is particularly Jesuit is identifying that joy with the presence of God. It is the attempt to find God in those stars, to know that the mere presence and beauty of stars are a very particular sort of "something" that answers Leibniz's famous query, "Why is there something instead of nothing?"

The second trait of Jesuit astronomy is that Jesuits are not satisfied with simply enjoying the view; we want to engage our minds as well as our hearts. And while of course I am thinking of the mathematical or scientific study of the phenomena, a Jesuit astronomy also involves the wonderful academic subjects of history and philosophy, photography and poetry.

The third trait? Ah, my Jesuit education also has taught me that every question can, like all of Gaul, be divided into three parts. So what might that third trait be? Education, of course. The third trait of Jesuit astronomy involves telling the whole world, from China to our high school classrooms, about what we have learned. Giving classes, giving talks, and writing books like this one.

We, nature and humanity, are both
children of the same Father.

FURTHER READING

Asimov, Isaac. *Nightfall and Other Stories.* Doubleday, 1969.

Bell, Jim. *Postcards from Mars: The First Photographer on the Red Planet.* Penguin Books, 2006.

Brown, Mike. *How I Killed Pluto and Why It Had It Coming.* Penguin Random House, 2012.

Chesterton, G. K. *Orthodoxy.* Dodd, Mead & Company, 1908.

Chinnici, Ileana. *Decoding the Stars: A Biography of Angelo Secchi, Jesuit and Scientist.* Brill, 2019.

Chinnici, Ileana, and Guy Consolmagno, eds. *Angelo Secchi and Nineteenth Century Science: The Multidisciplinary Contributions of a Pioneer and Innovator.* Vatican Observatory Publications, 2020.

Clarke, Arthur C. *Rendezvous with Rama.* Harcourt Brace Jovanovich, 1973.

Clement, Hal. *Mission of Gravity.* Pyramid Books, 1974.

Consolmagno, Guy, and Dan M. Davis. *Turn Left at Orion: Hundreds of Night Sky Objects to See in a Home Telescope—and How to Find Them.* Cambridge University Press, 2011.

Eliot, T. S. *The Rock.* Faber & Faber, 1934.

Francis. *Laudato si´: On Care for Our Common Home.* Vatican Press, 2015.

Francis of Assisi. *Laudato si´: Canticle of Brother Sun and Sister Moon.* Franciscan Institute Publications, 1976.

Herbert, Frank. *Dune.* Chilton Books, 1965.

Hopkins, Gerard Manley. *The Poems of Gerard Manley Hopkins.* Edited with notes by Robert Bridges. Oxford University Press, 1918.

Ignatius of Loyola. *Autobiography of St. Ignatius.* Translated by Joseph F. O'Callaghan. Fordham University Press, 1992.

Ignatius of Loyola. *The Spiritual Exercises of St. Ignatius.* Translated by Anthony Mottola. Doubleday, 1964.

John Paul II. *Fides et ratio: On the Relationship between Faith and Reason.* Vatican Press, 1998.

John Paul II. "Letter of His Holiness John Paul II to Reverend George V. Coyne, SJ." Vatican Press, 1988.

Kierkegaard, Søren. *Fear and Trembling.* Translated by Alastair Hannay. Penguin Books, 1985.

Levy, David H. *The Starlight Night: The Sky in the Writings of Shakespeare, Tennyson, and Hopkins.* Springer, 1997.

Lewis, C. S. *The Chronicles of Narnia.* HarperCollins, 1950–1956.

Lewis, C. S. *The Discarded Image: An Introduction to Medieval and Renaissance Literature.* Cambridge University Press, 1964.

Locher, Johann Georg. *Mathematical Disquisitions: The Booklet of Theses Immortalized by Galileo.* Translated by Christopher Graney. University of Notre Dame Press, 2017.

Martin, James. *The Jesuit Guide to (Almost) Everything: A Spirituality for Real Life.* HarperOne, 2010.

Niven, Larry. *Ringworld.* Random House Publishing Group, 1970.

O'Malley, J. W., and G. A. Bailey. *The Jesuits and the Arts.* Saint Joseph's University Press, 2005.

O'Malley, John. *The First Jesuits.* Harvard University Press, 1993.

Rey, H. A. *The Stars: A New Way to See Them.* Houghton Mifflin Harcourt, 1952.

Riccioli, Giovanni Battista. *Setting Aside All Authority: Giovanni Battista Riccioli and the Science against Copernicus in the Age of Galileo.* Translated by Christopher Graney. University of Notre Dame Press, 2015.

Sacks, Jonathan. *The Great Partnership: Science, Religion, and the Search for Meaning.* Schocken, 2011.

Schmitz, James. *The Demon Breed.* Ace, 1968.

Spufford, Francis. *Unapologetic: Why, Despite Everything, Christianity Can Still Make Surprising Emotional Sense.* HarperOne, 2013.

Stoeger, William. "Is Big Bang Cosmology in Conflict with Divine Creation?" in *The Heavens Proclaim.* Edited by G. Consolmagno. Our Sunday Visitor, 2009.

Tolkien, J. R. R. *The Lord of the Rings.* Allen & Unwin, 1954–1955.

Udías, Agustín, SJ. *Jesuit Contribution to Science, A History.* Springer, 2015.

Udías, Agustín, SJ. *Searching the Heavens and the Earth: The History of Jesuit Observatories.* Springer, 2003.

White, T. H. *The Once and Future King.* Collins, 1958.

BOOKS THAT WERE FORMATIVE IN MY LIFE

WHEN IT COMES TO LEARNING PRACTICAL ASTRONOMY, my favorite guide is H. A. Rey's lovely book *The Stars: A New Way to See Them.* (Rey also gave us the Curious George books.) It's a gentle and fun introduction to the constellations, which also includes a remarkably clear description of why the stars and planets move in the sky the way they do over the course of a year. This book has been in print since before I was born; I still have the copy I got when I was eleven years old, but I also have a newer edition that I reference when I give classes to students and adults. It's the book I took with me to Kenya, the one that showed me how to find the Southern Cross. It has served as an inspiration to generations of students who have gone on to become professional astronomers.

It's surprisingly hard to find good books for a popular audience that actually describe what it's like to walk around (or fly above) real planets. Although it is nearly twenty years old now, Jim Bell's *Postcards from Mars* provides a sense of what at least one planet looks like, close up. For a fun description of the life of a planetary scientist, I relate to Mike Brown's provocatively titled memoir *How I Killed Pluto and Why It Had It Coming.*

Since my road to science started with science fiction, it's appropriate to mention some of the science fiction books that got me there. In many of them, the settings in space are themselves important characters in the story: Arthur C. Clarke's *Rendezvous with Rama,* Hal Clement's *Mission of Gravity,* and Larry Niven's *Ringworld* immediately come to mind. Frank Herbert's *Dune* is the epitome of that genre. I have a particular soft spot for a modest novel by James Schmitz, *The Demon Breed*—its clever heroine's derring-do occurs within an intricate floating island on an ocean planet. In retrospect, I recognize that the desert planet of *Dune* and the water planet of *The Demon Breed* suffer from the same syndrome: they are one-note ecosystems, exaggerations of what is found already on Earth. Reality, seen in the bizarre array of planets we have discovered since then, is so much harder to invent!

For a good sense of who the Jesuits are, a classic text is *The First Jesuits* by John O'Malley, SJ. Not only does O'Malley describe the essential traits of the Jesuits and how the order came to be; in his style and choice of topics, he demonstrates how Jesuits today think about themselves and the world around them.

My discussion on the difference between the Big Bang and creation from nothing owe a great debt to the late William Stoeger, SJ, especially his chapter "Is Big Bang Cosmology in Conflict with Divine Creation?" in a book I edited, *The Heavens Proclaim*.

Along with C. S. Lewis and G. K. Chesterton, who taught me how to see my religion as a great adventure and how to write about it, one modern writer who helped me see the importance of going beyond clever arguments in talking about faith is Francis Spufford. His book *Unapologetic: Why, Despite Everything, Christianity Can Still Make Surprising Emotional Sense,* is certainly fun to read. But more, it reminds us that we human beings are guided by heart and head together. The adventure of Christianity is our adventure; adventures always come with challenges and setbacks; and most certainly, echoing *The Lord of the Rings*'s Bilbo Baggins at the Council of Elrond, we are always left wondering, *Don't adventures ever end?*

ACKNOWLEDGMENTS

The idea for this book, and its title, were suggested to me by Gary Jansen at Loyola Press, who a few years previously had edited the book I wrote with Paul Mueller, SJ, *Would You Baptize an Extraterrestrial?*. Apparently, that experience has not permanently scarred him. He's been a tremendous support throughout this effort. My editor Maura Poston at Loyola Press and my agent Gillian MacKenzie complete the team who made this book possible. Thanks to them both!

My remarkable friend Dennis McCarthy read through the text several times. One of the most widely read people I know, he was able to call me out on a number of places where I had made sloppy references to history, theology, or popular culture. And since we've known each other since high school, he has no hesitation in letting me know when I need to be caught! I could not have asked for a more perfect critic and guide.

Material in chapters "An Astronomer's Road to the Jesuits" and "Star Relics" first appeared, in Italian, in *Sole* (Perugia: Edizioni Frate Indovino, 2021). Some of "Astronomer's Road" is also adapted from one of my columns in the British weekly the *Tablet*.

The 2018 bicentennial of Angelo Secchi's birth led to the publication of a number of books about him and his science, most notably Dr. Ileana Chinnici's excellent biography *Decoding the Stars: A Biography of Angelo Secchi, Jesuit and Scientist* (Brill, 2019), and two compendiums of chapters by experts in the various fields that Secchi worked in: *Padre Angelo Secchi: La figura, le opere, l'astrofisica,* edited by S. Chicchi and R. Marcuccio (Nero Colore srl, 2021) and *Angelo Secchi and Nineteenth Century Science: The Multidisciplinary Contributions of a Pioneer and Innovator,* edited by I. Chinnici and G. Consolmagno (Springer Nature Switzerland AG, 2021). Material that I originally contributed to those books has been incorporated into the Secchi chapter.

Unless otherwise indicated, all photographs are by the author and all astrophotographs were taken by astronomers of the Vatican Observatory at either our Vatican Advanced Technology Telescope or other smaller telescopes.

Citations accompanying crater names are from the online *IAU Gazetteer of Planetary Nomenclature* (planetarynames.wr.usgs.gov). Citations accompanying asteroid names are from the *Minor Planet Center & Dictionary of Minor Planet Names*. Grateful thanks to Br. Robert J. Macke, SJ, who assembled the list of asteroids, and to Andrew Kassebaum, Gareth Williams, and Bob Trembley, who assisted him by contributing to this list.

#exemplum omnibus

MAUPAL
'19

A mural by the artist MAUPAL, or Mauro Pallotta, commissioned by the town leaders in Albano, next to the Vatican Observatory's headquarters, upon the occasion of Pope Francis's visit.

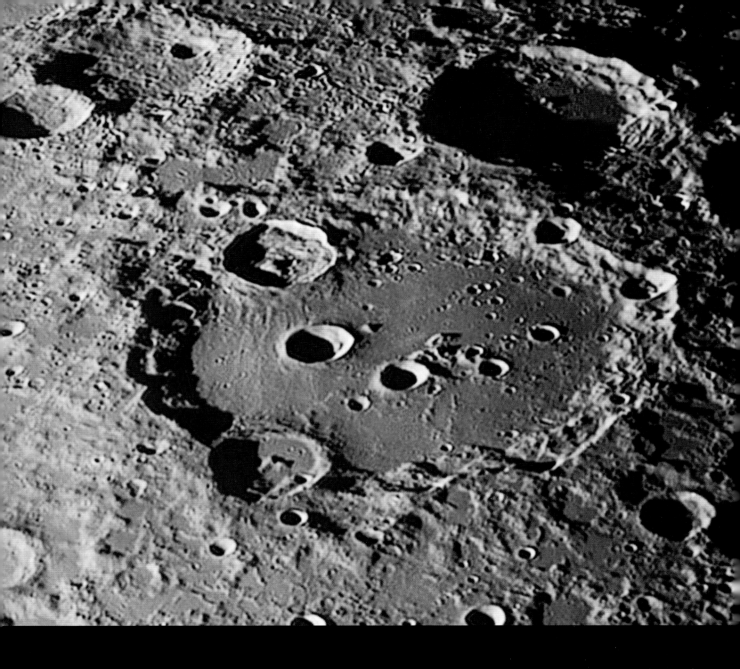

Crater Clavius, named for Christopher Clavius, SJ, and the site of the fictional Moon base in the film *2001: A Space Odyssey*; image by Vatican Observatory Foundation blogger Rik Hill. South is up.

APPENDIX

JESUIT NAMES ASSOCIATED WITH ASTEROIDS AND LUNAR CRATERS

IN 1652, GIOVANNI BATTISTI RICCIOLI, SJ, published *Almagestum novum*, which contained a Moon map that devised the system of nomenclature we use to this day. In particular, he began the tradition of naming craters for important scientists. Notably, the most prominent craters on the Moon were named Copernicus and Tycho, whose competing cosmologies were quite controversial in his time. (Galileo and Kepler also received craters.) Among the others getting crater names were two dozen fellow Jesuits, including Riccioli's student Francesco Maria Grimaldi, SJ (1618–1663), and Riccioli himself.

In 1935, the International Astronomical Union took over the naming of solar system features, adopting the names assigned by Riccioli and others up to that time. Today, crater names are approved by the IAU Working Group on Planetary System Nomenclature, with new names assigned only when a scientific need can be demonstrated for adding a name to a particular feature, such as when someone is writing a paper about an unnamed crater and wants a name for it. The current rules insist that the person so honored has been deceased at least three years.

Meanwhile, asteroids were originally thought to be few in number and similar to planets, so like the planets they were just given the names of Roman gods, such as Ceres and Vesta. By earliest tradition, the discoverer of the asteroid gets to choose the name, now subject to approval by the IAU Working Group for Small Body Nomenclature. Today asteroids are numbered, generally in the order in which their orbits are approved as well defined, and then they are given a name. Thus, the first asteroid discovered (now considered a dwarf planet) is designated 1 Ceres.

By the mid-1800s, so many asteroids had been discovered that nonclassical names were beginning to be used for them. Today we know of more than a million asteroids with reliable orbits that are eligible for names. According to the IAU, asteroid names must follow these rules: sixteen characters or less in length, preferably one word, pronounceable (in some language), nonoffensive, and not too similar to an existing name of another asteroid or a natural planetary satellite. The names of individuals or events principally known for political or military activities are unsuitable until one hundred years after the death of the individual or the occurrence of the event. In addition, names of pet animals are discouraged.

Here, in alphabetical order by their name, are the Jesuits who have craters or asteroids bearing names in their honor, as well as a diagram of the locations of some of these asteroids. This list is current to December 1, 2023. The descriptions are based on the official IAU designations, which are much longer for asteroids than for craters and not always guaranteed to be completely accurate.

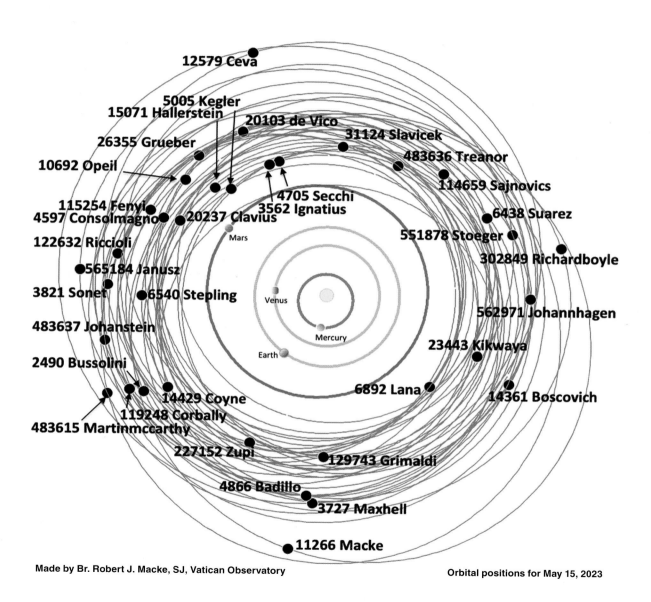

Made by Br. Robert J. Macke, SJ, Vatican Observatory

Orbital positions for May 15, 2023

The locations of asteroids named for Jesuits.

Asteroid 4866 Badillo: Victor L. Badillo, SJ (1930–2014); popularized astronomy in the Philippines for more than three decades, inspiring countless Filipino astronomers. Ordained in 1965, he directed the Jesuit-run Manila Observatory in Quezon City and served as president of the Philippine Astronomical Society from 1972 to 1990.

Crater Bettinus: Mario Bettinus, SJ (1582–1657); Italian mathematician and astronomer.

Crater Billy: Jacques de Billy, SJ (1602–1679); French mathematician.

Crater Blancanus: Giuseppe Biancani, SJ (1566–1624); Italian mathematician and astronomer.

Crater Boscovich; asteroid 14361 Boscovich: Ruggiero Giuseppe Boscovich, SJ (1711–1787); Jesuit professor of mathematics and philosophy at Rome and Pavia. For some years lived in Paris and later Milan, where he founded the Brera Astronomical Observatory. He wrote on the determination of orbits of comets, mathematics, and optics.

Asteroid 302849 Richardboyle: Richard Boyle, SJ (b. 1943); astronomer at the Vatican Observatory. His work, specialized in high-precision photometry of stars and stellar clusters and often using Vilnius system filters, has application ranging from asteroseismology to the discoveries of asteroids with the Vatican Advanced Technology Telescope.

Asteroid 2490 Bussolini: Juan Antonio Bussolini, SJ (1905–1966); Argentine Jesuit and solar physicist and director of the Observatorio de Fisica Cosmica de San Miguel. He was a noted supporter of the Félix Aguilar Observatory.

Crater Cabeus: Niccolo Cabeo, SJ (1586–1650); Italian astronomer.

Asteroid 620307 Casanovas: Juan Casanovas, SJ (1929–2013); solar astronomer who helped establish solar observatories in the Canary Islands and later served as an astronomer and librarian at the Vatican Astronomical Observatory in Castel Gandolfo. He was also interested in the history of astronomy, in particular calendars.

Asteroid 12579 Ceva: The brothers Giovanni (1647–1734) and Tommaso (1648–1737) Ceva were Italian mathematicians interested in geometry and physics. Giovanni was in the service of the Duke of Mantua. Tommaso was a Jesuit and professor of mathematics in the Jesuit college at Milan and was also known as a poet.

Crater Clavius; asteroid 20237 Clavius: Christopher Clavius, SJ (1538–1612); German mathematician and astronomer. He figured out where to place the leap years in the Gregorian calendar. Pope Gregory XIII revised the Julian calendar with the assistance of Clavius.

Asteroid 4597 Consolmagno: Guy Consolmagno, SJ (b. 1952); studied the origins of eucritic meteorites. As the curator of the Vatican meteorite collection, Guy's more recent efforts have focused on determining the densities and porosities of meteorites and making comparisons with the densities of minor planets.

Asteroid 119248 Corbally: Christopher J. Corbally, SJ (b. 1946); astronomer. He continues a long career in astronomy, where his contributions have included areas of multiple stellar systems, stellar spectral classification, galactic structure, star formation, and telescope technology.

Asteroid 14429 Coyne: George Coyne, SJ (1933–2020); astronomer at the Vatican Observatory (1969–2006) and its director (1978–2006). He helped with the completion of the large Vatican telescope on Mount Graham, Arizona. His polarimetric studies centered on cataclysmic variables, among other subjects.

Crater Cysatus: Jean-Baptiste Cysat, SJ (1587–1657); Swiss mathematician and astronomer.

Crater De Vico; asteroid 20103 de Vico: Francesco de Vico, SJ (1805–1848); best known as the discoverer of seven comets, including 54P and 122P. He was the director of the Observatory of Collegio Romano in Rome from 1838 to 1848. De Vico was born in the city of Macerata, and the Osservatorio di Monte D'Aria di Serrapetrona is dedicated to him.

Crater Fényi; asteroid 115254 Fényi: Gyula Fényi, SJ (1845–1927); Hungarian astronomer. A most prolific observer of solar phenomena, Fényi is noted for his thirty-two-year continuous series of prominence observations with the same instrument. He was the first to demonstrate a correlation between the numbers of solar prominences and sunspots.

Crater Furnerius: Georges Fournier, SJ (1595–1652); French mathematician.

Crater Grimaldi; asteroid 129743 Grimaldi: Francesco Maria Grimaldi, SJ (1618–1663); one of the first telescopic observers of the Moon. He wrote *Physico-mathesis de lumine, coloribus, et iride* (Physicomathematical Studies of Light, Colors, and the Rainbow), the first treatise on the wave nature of light.

Asteroid 26355 Grueber: Johannes Grueber, SJ (1623–1680); priest, missionary, mathematician, and astronomer at the Chinese imperial court from 1659 to 1661. He returned to Europe from China by the overland route and published the very first travelogue describing Tibet.

Crater Gruemberger: Christoph Gruemberger, SJ (1561–1636), name more commonly rendered Grienberger; Austrian astronomer.

Crater Hagen; asteroid 562971 Johannhagen: Johann Georg Hagen, SJ (1847–1930); Austrian-American astronomer and priest. He was director at the Georgetown University Observatory (1888–1906) and the Vatican Observatory (1906–1930). Hagen devised several ingenious experiments at the Vatican to demonstrate Earth's rotation, directly confirming the theories of Copernicus and Galileo.

Asteroid 15071 Hallerstein: Ferdinand Avguštin Hallerstein, SJ (1703–1774); Ljubljana-born missionary in China, known for his work in astronomy, mathematics, and cartography. As president of the Chinese Imperial Bureau of Astronomy, he prepared an accurate catalog of 3,083 stars and discovered and carefully measured comet C/1748 H1.

Crater Hell; asteroid 3727 Maxhell: Maximilian Hell, SJ (1720–1792); determined the solar parallax from his observations of the transit of Venus in 1769. He was appointed director of the Imperial Observatory in Vienna in 1756. He prepared and published an important series of astronomical ephemerides.

Asteroid 3562 Ignatius: Ignatius of Loyola (1491–1556), founder of the Jesuits (note also asteroid 3589 Loyola, named for the town in Spain, birthplace of Ignatius).

Asteroid 565184 Janusz: Robert M. Janusz, SJ (b. 1964); Polish priest, philosopher, and physicist noted for his study of star clusters and interstellar matter using data taken in the Vilnius System at the Vatican Advanced Technology Telescope. His philosophical work is noted for its discussion of field theory, computer science, and the mathematical nature of the universe.

Asteroid 5005 Kegler: Ignaz Kögler, SJ (1680–1746); German missionary in Qing-dyansty China, where he earned favor in the imperial court and was named president of the Chinese mathematical astronomical tribunal.

Asteroid 23443 Kikwaya: Jean-Baptiste Kikwaya Eluo, SJ (b. 1965); from Democratic Republic of Congo and staff astronomer at the Vatican Observatory. Using optical meteor measurements, he estimates the bulk densities of smaller meteoroids through numerical ablation models.

Crater Kircher: Athanasius Kircher, SJ (1602–1680); German humanitarian.

Crater Kugler: Franz Xaver Kugler, SJ (1862–1929); German-Babylonian chronologist.

Asteroid 6892 Lana: Francesco Lana de Terzi, SJ (1631–1687); Italian professor of physics and mathematics who first explored the concept for a vacuum airship, bringing human flight into the field of science. He also originated the idea and concept of an alphabet for the blind, developed later by Louis Braille.

Asteroid 11266 Macke: Robert J. Macke, SJ (b. 1974); research scientist and meteorite curator at the Vatican Observatory, whose fundamental contributions include studying the relationship between shock state and porosity in carbonaceous chondrites.

Asteroid 53053 Sabinomaffeo: Sabino Maffeo, SJ (b. 1922); educator in the Massimiliano Massimo Jesuit college in Rome, provincial of the Roman Province of the Society of Jesus, and technical director of Vatican Radio. He was later appointed vice director and community superior of the Vatican Observatory.

Crater Malapert: Charles Malapert, SJ (1581–1630); Belgian astronomer, mathematician, and philosopher.

Crater C. Mayer: Christian Mayer, SJ (1719–1783); German astronomer, mathematician, and physicist.

Asteroid 483615 Martinmccarthy: Martin F. McCarthy, SJ (1923–2010); American astronomer noted for his study of carbon stars. At the Vatican Observatory from 1958 to 1999, he served as a key figure in the transition to the world of modern research. In 1986 he founded the Vatican Observatory Summer Schools in Astrophysics, an initiative that has become world renowned.

Crater McNally: Paul Aloysius McNally, SJ (1890–1955); American astronomer.

Crater Moretus: Théodor Moret, SJ (1602–1667); Belgian mathematician.

Asteroid 10692 Opeil: Cyril P. Opeil, SJ (b. 1960); professor at Boston College studying the thermal properties of meteorites to improve understanding of orbital and rotational changes caused by the reradiation of solar flux.

Crater Petavius: Denis Pétau, SJ (1583–1652); French chronologist and astronomer.

Crater Riccioli; asteroid 122632 Riccioli: Giovanni Battista Riccioli, SJ (1598–1671); one of the first telescopic observers of the Moon. He was author of the *Almagestum novum*, which contains a lunar map still used today.

Crater Riccius: Matteo Ricci, SJ (1552–1610); Italian mathematician and geographer.

Asteroid 114659 Sajnovics: János Sajnovics, SJ (1733–1785); Hungarian linguist. He is best known for his pioneering work in comparative linguistics, particularly his systematic demonstration of the relationship between Sami and Hungarian. Sajnovics was a pupil of the astronomer and mathematician Maximilian Hell.

Crater Scheiner: Christoph Scheiner, SJ (1573–1650); German astronomer.

Crater Schomberger: Georg Schomberger, SJ (1597–1645); Austrian astronomer and mathematician.

Crater Secchi; asteroid 4705 Secchi: Angelo Secchi, SJ (1818–1878); Italian astronomer and director of the observatory of the Collegio Romano in Rome from 1848 to 1878. Famous for his work on stellar spectroscopy, he made the first spectroscopic survey of the heavens, and his classification scheme divided the spectra of the stars into four groups. Secchi also made an extensive study of solar phenomena and was a co-founder of the Società degli Spettroscopisti Italiani, now the Società Astronomica Italiana.

Crater Simpelius: Hugh Sempill, SJ (1589–1654); Scottish mathematician.

Crater Sirsalis: Gerolamo Sersale, SJ (1584–1654); Italian astronomer.

Asteroid 31124 Slavíček: Karel Slavíček, SJ (1678–1735); missionary and scientist and the first Czech sinologist. Together with Ignatius Kegler he went to China in 1716. He worked on astronomy, mathematics, and music and prepared maps of Beijing in 1718 and 1728.

Asteroid 3821 Sonet: Jean Sonet, SJ (1908–1987); Belgian specialist in Romance languages, and a professor and later rector (1953–1958) of the University of Namur. From

1958 to his death, he was vice rector of the Catholic University of Córdoba (Argentina), where the discoverer met him.

Crater Stein; asteroid 483637 Johanstein: Johan Stein, SJ (1871–1951); Dutch astronomer and a student of H. A. Lorentz, noted for his work on variable stars. He served as director of the Vatican Observatory from 1930 until his death. During World War II he opened the observatory to twelve thousand refugees. Queen Juliana made him a Knight of the Lion of Holland.

Asteroid 6540 Stepling: Joseph Stepling, SJ (1716–1778); founder and the first director of the astronomical observatory at the Jesuit college in Prague called Klementinum (1751). He was also known for introducing Newtonian physics to Prague.

Asteroid 551878 Stoeger: William R. Stoeger, SJ (1943–2014); American priest at the Vatican Observatory who developed ways to test different mathematical formalisms of cosmology by observation. He coedited a notable series of academic conference proceedings on science and theology with colleagues at the Center for Theology and the Natural Sciences in Berkeley, California.

Asteroid 6438 Suarez: Buenaventura Suárez, SJ (1679–1750); pioneer native astronomer of the Río de la Plata. He established the first observatory of the region in San Cosme y Damián, Paraguay, where he made observations of eclipses of Jupiter's satellites. His *Lunario de un siglo* included computations of eclipses and lunar phases.

Crater Tacquet: André Tacquet, SJ (1612–1660); Belgian mathematician.

Crater Tannerus: Adam Tanner, SJ (1572–1632); Austrian mathematician.

Asteroid 483636 Treanor: Patrick Treanor, SJ (1920–1978); English astronomer noted for research on polarized light from stars and the interstellar medium. At Oxford, he won the Johnson Memorial Prize for his doctoral thesis on interference phenomena. He served as director of the Vatican Observatory from 1970 until his death.

Crater Zucchius: Niccoló Zucchi, SJ (1586–1670); Italian mathematician and astronomer.

Crater Zupus; asteroid 227152 Zupi: Giovanni Battista Zupi (or Zupus), SJ (c. 1589–1650); Italian astronomer, mathematician, and priest. In 1639, he was the first person to discover that the planet Mercury showed orbital phases, like those of the Moon and Venus.

PERMISSIONS

The adventure of Christianity is our adventure; adventures always come with challenges and setbacks; and most certainly, echoing *The Lord of the Rings's* Bilbo Baggins at the Council of Elrond, we are always left wondering,

Don't adventures ever end?

ART ACKNOWLEDGMENTS

Cover: (t) fotograzia/Moment/Getty Images. (c) Stefano Spaziani; (bg) Photo courtesy of Roelof de Jong. **Interior: i** fotograzia/Moment/Getty Images; Photo courtesy of Roelof de Jong. **iii** fotograzia/Moment/Getty Images. **vi** Photo courtesy of Roelof de Jong. **viii** Vatican Observatory licensed under CC. **x** Stefano Spaziani. **xii** (bg) Natalya Bosyak/iStock/Getty Images; From the book Ignace de Loyola, Une vie en vingt tableaux/Igantius von Loyola, Ein Leben in zwansig Bildern [in French and German] (Basel: Friedrich Reinhardt Verlag), 2014/Public Domain. **xv** View of Earth over Moon's horizon taken from Apollo 11 spacecraft courtesy of NASA. NASA ID: AS11-44-6552. Date Created: 1969-07-16. **xvi** Claudio Costa/Vatican Observatory licensed under CC. **2** Photo by author Guy Consolmagno, SJ; (bg) Natalya Bosyak/iStock/Getty Images. **5** Brucker, Consolmagno, Romanishin, and Tegler/Vatican Observatory licensed under CC. **6** Vatican Observatory licensed under CC. **8** Claudio Costa/Vatican Observatory licensed under CC. **10** Brucker, Consolmagno, Romanishin, and Tegler/Vatican Observatory licensed under CC. **12** Claudio Costa/Vatican Observatory licensed under CC. **15** Photo by author Guy Consolmagno, SJ. **16** Vatican Observatory licensed under CC. **17** Matt Nelson/Vatican Observatory licensed under CC. **18** Br. Jonathan Stott, SJ/Vatican Observatory licensed under CC. **20** Karl Treusch/Vatican Observatory licensed under CC. **23** Vatican Observatory licensed under CC. **24** Photo courtesy of author Guy Consolmagno, SJ. **29** Photo courtesy of author Guy Consolmagno, SJ. **30** Photo courtesy of author Guy Consolmagno, SJ. **31** Photo courtesy of author Guy Consolmagno, SJ. **35** Vatican Observatory licensed under CC. **39** Natalya Bosyak/iStock/Getty Images. **40** Br. Jonathan Stott, SJ/Vatican Observatory licensed under CC. **44** View of Atlantis's Payload Bay courtesy of NASA. NASA ID: s125e012372. Date Created: 2009-05-20. **48** Manny Carriera, SJ/Vatican Observatory licensed under CC. **57** (l) Image courtesy of Dan Davis; (r) Vatican Observatory licensed under CC. **58)** (l) Image courtesy of Dan Davis; (r) Vatican Observatory licensed under CC. **61** (l) Image courtesy of Dan Davis; (r) Vatican Observatory licensed under CC . **62** (l) Image courtesy of Dan Davis; (r) Vatican Observatory licensed under CC. **63** (l) Image courtesy of Dan Davis; (r) Vatican Observatory licensed under CC. **64** Photo by author Guy Consolmagno, SJ. **66** Photo by author Guy Consolmagno, SJ. **73** Br. Robert J. Macke, SJ/Vatican Observatory licensed under CC. **74** Vatican Observatory licensed under CC. **77** Photo by author Guy Consolmagno, SJ. **78** Photo by author Guy Consolmagno, SJ. **83** Photo by author Guy Consolmagno, SJ. **84** Photo by author Guy Consolmagno, SJ. **87** Marko Valentini/Vatican Observatory licensed under CC. **95** Br. Guy Consolmagno, SJ/Vatican Observatory

ABOUT THE AUTHOR

BROTHER GUY CONSOLMAGNO, SJ is the Director of the Vatican Observatory. A native of Detroit, Michigan, he earned undergraduate and master's degrees from Massachusetts Institute of Technology (MIT), and a Ph. D. in Planetary Science from the University of Arizona. He was a research fellow at Harvard and MIT, served in the US Peace Corps (Kenya), and taught university physics at Lafayette College before entering the Jesuits in 1989. At the Vatican Observatory since 1993, in 2015 Pope Francis appointed Dr. Consolmagno director of the Vatican Observatory.

Br. Guy's research explores connections between meteorites, asteroids, and the evolution of small solar system bodies. In 2000, the International Astronomical Union named asteroid 4597 Consolmagno in recognition of his work. In 2014 he received the Carl Sagan Medal from the American Astronomical Society Division for Planetary Sciences for excellence in public communication in planetary sciences.

Along with more than 250 scientific publications, Br. Guy is the author of a number of popular astronomy books.

The domes of the Vatican Observatory beside the dome of St. Thomas of Villanova in Castel Gandolfo; photo by Judith Britt.

The emission nebula NCG 7000 in the constellation Cygnus is popularly known as the
North America Nebula. Imaged at Vatican Advanced Technology Telescope.

This galaxy in Virgo, M104, is sometimes called the Sombrero; dust blocks out the light from the edge of its disk. This image was taken at Vatican Advanced Technology Telescope.

The Vatican Observatory is a centuries-old gathering place where scientists and scholars from all around the world conduct pioneering interdisciplinary research to further humankind's understanding of the universe.

See https://www.vaticanobservatory.org/ for more information.

EXPLORE THE UNIVERSE WITH THE POPE'S ASTRONOMER

JOURNEY THROUGH THE AWE-INSPIRING COSMOS with celebrated Jesuit astronomer Br. Guy Consolmagno in his masterful exploration of the universe and our place in it.

To turn each page of *A Jesuit's Guide to the Stars* is to appreciate anew the graceful alignment of God and science. Br. Guy leads the way as we discover the unbroken connection between Scripture, tradition, and humankind's quest to see God in all things—even, or maybe especially, in the stars.

Brought to life with more than 70 stunning images of the cosmos, Br. Guy seamlessly integrates his studies and knowledge as a scientist with his personal journey as a man of faith. His uniquely inspiring story connects the heavens to the earth, the past to the present, and faith to science.

Whether you are a curious stargazer, an amateur astronomer, or someone seeking meaning in a universe that can seem impersonal, *A Jesuit's Guide to the Stars* will show you that the cosmos is a beautiful subject to contemplate and a joyful way to find God.

BROTHER GUY CONSOLMAGNO, SJ, is a native of Detroit, Michigan. He earned his undergraduate and master's degrees from Massachusetts Institute of Technology (MIT) and a PhD in planetary science from the University of Arizona. He entered the Jesuits in 1989. He has served at the Vatican Observatory since 1993, and in 2015 Pope Francis appointed him director. He is the author of *God's Mechanics* and co-author of *Would You Baptize an Extraterrestrial?* and *Turn Left at Orion*.

LOYOLA PRESS.
A JESUIT MINISTRY
www.loyolapress.com

ISBN: 978-0-8294-5573-1 $24.99 U.S.

52499

9 780829 455731